Adult Stem Cells in Aging, Diseases and Cancer

Else Kröner-Fresenius Symposia

Vol. 5

Series Editor

S. Pahernik Heidelberg

Adult Stem Cells in Aging, Diseases and Cancer

Volume Editor

Karl Lenhard Rudolph Jena

19 figures in color, 2015

Basel · Freiburg · Paris · London · New York · Chennai · New Delhi · Bangkok · Beijing · Shanghai · Tokyo · Kuala Lumpur · Singapore · Sydney

Karl Lenhard Rudolph
Leibniz Institute for Age Research
Fritz Lipmann Institute e.V. (FLI)
Beutenbergstrasse 11
DE–07745 Jena (Germany)

This book is sponsored by the Else Kröner-Fresenius-Stiftung.

Else Kröner-Fresenius Symposium on Adult Stem Cells in Aging, Diseases, and Cancer (2013 : Eisenach, Germany), author.
Adult stem cells in aging, diseases, and cancer / volume editor, Karl Lenhard Rudolph.
p. ; cm. -- (Else Kröner-Fresenius symposia, ISSN 1663-0114 ; vol. 5)
Includes bibliographical references and indexes.
ISBN 978-3-318-02731-0 (hard cover : alk. paper) -- ISBN 978-3-318-02732-7 (eISBN)
I. Rudolph, K. Lenhard, editor. II. Title. III. Series: Else Kröner-Fresenius symposia ; v. 5. 1663-0114
[DNLM: 1. Adult Stem Cells--physiology--Congresses. 2. Cell Aging--physiology--Congresses. 3. DNA Damage--physiology--Congresses. 4. Neoplasms--genetics--Congresses. 5. Telomere--genetics--Congresses. QU 325]
QH588.S83
616.02'774--dc23
2014040523

Bibliographic Indices. This publication is listed in bibliographic services, including Current Contents®.

Cover art by Maren Blaschke

www.karger.com
Printed in Germany on acid-free and non-aging paper (ISO 9706) by Kraft Druck, Ettlingen
ISSN 1663–0114
ISBN 978–3–318–02731–0
eISBN 978–3–318–02732–7

Contents

Preface

Adult Stem Cells in Aging, Diseases and Cancer

This book series features the proceedings of the Else Kröner-Fresenius Symposia, which cover clinically relevant topics at the forefront of biomedical research. The meetings should give experts the opportunity to discuss the most recent findings in evolving fields of biomedicine and outline future research strategies.

Today's research is characterized by the accelerated generation of biomedical data, the increasingly interdisciplinary and translational nature of biomedical science as well as the efforts to integrate the data into complex biological systems. These developments emphasize the need for new forums of discussion.

The understanding of stem cell aging in the context of pathological tissue and malignant tumors is highly relevant for future developments in biomedicine. Physiology, modulation and pathology might be promising to open new perspectives to restore organ function lost by disease or physical trauma or to overcome pathophysiological conditions.

In this context, Prof. Dr. Karl Lenhard Rudolph organized the 5th Else Kröner-Fresenius Symposium held at the famous Wartburg Castle near Eisenach in May 2013, a distinguished meeting which focused on adult stem cells in aging, diseases and cancer. The Else Kröner-Fresenius Symposium with world-renowned experts discussed current advances in stem cell aging and illustrated new perspectives from the diverse fields of stem cell research, and how they could be integrated into clinical practice, in particular with respect to inflammatory disease and cancer.

The Else Kröner-Fresenius-Stiftung thanks Prof. Dr. Rudolph for his inspiring scientific work and the outstanding personal input in organizing the 5th Else Kröner-Fresenius Symposium with his team.

The Else Kröner-Fresenius-Stiftung

In 1983, Else Kröner (1925–1988) founded the Else Kröner-Fresenius-Stiftung, a nonprofit foundation dedicated to promoting medical science, supporting medical education and providing humanitarian aid.

Else Kröner, née Fernau, was born on May 15, 1925, in Frankfurt am Main, Germany. Her father died when she was only 3 years old. After his death, she lived with her mother in the home of Dr. Eduard Fresenius, a pharmacist and owner of the Hirsch Pharmacy in Frankfurt, who founded the pharmaceutical company Fresenius in 1912. Dr. Eduard Fresenius, whose marriage was childless, took care of Else Fernau. In 1944, she started an internship at the Hirsch Pharmacy and decided to study pharmacy, which was supported by Dr. Eduard Fresenius. In 1946, Dr. Eduard Fresenius died unexpectedly. At this time, Else Fernau had not completed her pharmaceutical education. However, Dr. Eduard Fresenius bequeathed the Hirsch Pharmacy and the Fresenius company to her.

At the age of 21 years, Else Fernau decided, against the advice of many, to take responsibility for the Hirsch Pharmacy and the Fresenius company, which were then experiencing severe financial difficulties during the post-war years. Of the original 400 employees, all but 30 had to be laid off. In her efforts to ensure the survival and force the re-expansion of the company, she was later supported by her husband, Hans Kröner. The company was progressively rebuilt, and the need to maintain it determined all activities.

These important and far-sighted entrepreneurial business decisions in the 1950s and 1960s ensured the successful future development of the company. Decades of growth followed, in particular within the field of dialysis, nutrition and intensive care, leading to an internationally competitive enterprise and market leader in special areas of health care.

Else Kröner led the company until 1981. After the transformation of Fresenius into a stock company, she remained Chairwoman of the Supervisory Board until her death on June 5, 1988. From 1981 to 1992 her husband Hans Kröner led the company as CEO. Thereafter, he significantly shaped the policy of the Else Kröner-Fresenius-Stiftung, of which he was chairman of the board from 1995 to 2005.

Today, the Fresenius group, of which the Else Kröner-Fresenius-Stiftung is the leading share holder, is an international healthcare conglomerate with products and services for dialysis, hospital, and medical care of patients. The Fresenius group currently employs over 200,000 people in more than 100 countries and generates annual sales of over EUR 20 billion.

Else Kröner entrusted nearly her entire property to the foundation. She laid down that its financial resources should be employed to promote medical science, advance health care and provide humanitarian aid. It is in accordance with her vision and requirements that the Else Kröner-Fresenius-Stiftung continues to put the founder's fortune at the service of nonprofit projects and objectives. The symposia are published as part of the foundation's commitment to the advancement of medical research and treatment.

Sascha Pahernik, Heidelberg
Member of the Scientific Committee of the Else Kröner-Fresenius-Stiftung
Series Editor

Rudolph KL (ed): Adult Stem Cells in Aging, Diseases and Cancer.
Else Kröner-Fresenius Symp. Basel, Karger, 2015, vol 5, pp 1–2 (DOI: 10.1159/000366564)

Introduction

Karl Lenhard Rudolph

Leibniz Institute for Age Research, Fritz Lipmann Institute, Jena, Germany

Stem cells are present in almost all tissues of the adult human body. Aging-induced impairments in the functionality of stem cells in adult tissue turn out to be a major factor contributing to impairments in maintenance and regeneration of aging tissues, as well as to the development of aging-associated diseases. During the 5th Else Kröner-Fresenius Symposium, leading experts in

the field discussed novel concepts and unpublished results on molecular mechanisms of stem cell aging and their implications for the development of aging-associated dysfunction and diseases. The Symposium acted at the forefront of this emerging research field, which will not only change our understanding of disease evolution during aging but will also have a strong impact on the future development of molecular therapies aiming to increase health in the elderly.

Prof. Dr. med. K.L. Rudolph
Leibniz Institute for Age Research, Fritz Lipmann Institute e.V. (FLI)
Beutenbergstrasse 11
DE–07745 Jena (Germany)
E-Mail klrudolph@fli-leibniz.de

Rudolph KL (ed): Adult Stem Cells in Aging, Diseases and Cancer.
Else Kröner-Fresenius Symp. Basel, Karger, 2015, vol 5, pp 3–17 (DOI: 10.1159/000366563)

Speakers at the Symposium

Steven Artandi, MD, PhD

S.A. is a Professor in the Departments of Medicine and Biochemistry at Stanford University School of Medicine. He received his MD and PhD degrees from Columbia University and completed residency in internal medicine at Massachusetts General Hospital and a fellowship in oncology at the Dana-Farber Cancer Institute at Harvard Medical School. S.A. completed his postdoctoral studies in the laboratory of Dr. Ron DePinho at Harvard University and joined the faculty at Stanford in 2000. S.A.'s laboratory employs a broad interdisciplinary approach to understand cancer, stem cell function, and aging, including the use of mouse models, biochemistry, cell biology, proteomics, and genomics. His laboratory has pioneered the finding that telomerase has a direct role in stem cell regulation and cancer, independent of its telomere-elongating function. Through the generation and in-depth analysis of inducible telomerase mice, he revealed the unanticipated finding that telomerase activates quiescent stem cells in the skin. This function of telomerase does not involve its reverse transcriptase function; instead, S.A.'s laboratory has found that telomerase modulates Wnt signaling pathway, one of the most important circuits in self-renewal, stem cell regulation, and cancer. S.A. is a leader in the biochemical dissection of the human telomerase ribonucleoprotein complex, which has revealed new steps in telomerase assembly, trafficking and recruitment to telomeres. His laboratory has identified novel ATPases in telomerase assembly as well as a new component of the telomerase holoenzyme TCAB1, which is critical for the function of telomerase in human cells. His laboratory studies the stem cell disease dyskeratosis congenita using patient-derived induced pluripotent stem cells to understand how telomere shortening affects human stem cell populations. S.A. was elected as a fellow of the American Association for the Advancement of Science. He was elected member of the American Society for Clinical Investigation, and he serves on the editorial boards of Stem Cells and Molecular Cancer Research. He is the associate director of the Paul F. Glenn Laboratories for the Biology of Aging at Stanford, an institute to support and foster aging research at Stanford.

Axel Behrens, PhD

A.B. is senior scientist at the London Research Institute, Principal Research Fellow at King's College, London, and Honorary Professor at London University College, London (http://www.london-research-institute.org.uk/research/axel-behrens). He received his PhD degree in Molecular Genetics from Vienna University and completed his postdoctoral training in the University Hospital in Zurich (Switzerland). A.B.'s lab focuses on the biology of adult stem cells and cancer. His main contributions include (1) a mechanistic understanding of the JNK MAP kinase pathway, (2) the identification of the molecular connection between the Ras oncogene and activation of the AP-1 transcription factor, and (3) the characterization of a novel noncanonical arm of the ATM signaling pathway.

Cédric Blanpain, MD, PhD

C.B. is an MD/PhD and is studying the role of stem cells (SCs) during development, homeostasis, and cancer. As a postdoc in the laboratory of Elaine Fuchs, The Rockefeller University, he developed a new method to isolate hair follicle SCs and demonstrated their multipotency. He defined the role of Wnt and Notch signaling pathways in regulating epidermal SCs.

C.B. is Full Professor and WELBIO investigator at the Université Libre de Bruxelles. His lab has demonstrated the key role of Mesp1 during the specification of cardiovascular progenitors, identified the cellular origin of Merkel cells as well as the two most frequent skin cancers, and uncovered the mechanisms by which hair follicle SCs resist DNA damage-induced cell death. He also recently identified new SC populations in the mammary and the prostate epithelia. He developed new methods to study cancer stem cells by lineage tracing and define the mechanisms regulating their function.

C.B. is a member of the editorial board of *Stem Cells*, the *Journal of Cell Biology*, *Cell Reports*, *Stem Cell Reports*, *Development* and the *EMBO Journal*. C.B. received the EMBO Young Investigator award, the young investigator award of the ISSCR 2012, the Liliane Bettencourt award in life sciences 2012, and has been EMBO member since 2012.

Andrew Brack, PhD

A.B. is an Assistant Professor in the Department of Medicine at Harvard Medical School, a Principal Faculty at the Harvard Stem Cell Institute, and a Faculty member at Harvard Medical School's Division of Biological and Biomedical Sciences. He is also an Assistant Biologist at Massachusetts General Hospital's Center for Regenerative Medicine. A.B. received his BSc in Human Physiology from Manchester Metropolitan University, UK, in 1997 and his PhD in Molecular Biology and Biophysics from King's College London in 2001, where he also served as a postdoctoral fellow conducting research in the area of muscle stem cell biology. He came to America and continued his postdoctoral research in this field at Stanford University before joining Harvard Medical School and the Center for Regenerative Medicine as a lecturer and independent investigator in 2008.

A.B.'s current research interests lie at the interface of adult stem cell biology and tissue regeneration (www.bracklab.com). His lab focuses on the molecular pathways that control cell fate decisions of the adult muscle stem cell (the satellite cell) to effectively regenerate adult skeletal muscle. The Brack lab uses genetic strategies to deconstruct the communication between the niche and the muscle stem cell to investigate cell fate decisions during regeneration and aging. A.B. hopes that the lab's work on muscle stem cell regulation will lead to novel discoveries in stem cell regulation, their demise during aging and hopefully targeted strategies for therapeutic advantage.

Anne Brunet, PhD

A.B. is an Associate Professor in the Department of Genetics at Stanford University. A.B. obtained her BSc from the Ecole Normale Supérieure in Paris, France and her PhD from the University of Nice, France. She did her postdoctoral research training in Dr. Michael Greenberg's lab at Harvard Medical School. A.B. is interested in the molecular mechanisms of aging and longevity, with a particular emphasis on the nervous system. Her lab studies the molecular mechanism of action of known longevity genes, including FOXO transcription factors, in mammalian cells and organisms. She is particularly interested in the role of longevity genes in neural stem cells during aging. Another goal of the Brunet lab is to discover novel genes and processes regulating longevity using two model systems, the invertebrate *Caenorhabditis elegans* and an extremely short-lived vertebrate, the African killifish *Nothobranchius furzeri*. A.B. has received several grants from the National Institute on Aging to study the importance of FOXO transcription factors in aging neural stem cells, the molecular mechanisms of dietary restriction, and to develop genetic tools for the short-lived fish *N. furzeri*. She has published over 50 peer-reviewed papers, reviews, and book chapters. She has received a number of awards, including the Pfizer/AFAR Innovations in Aging Research Award, a Junior Investigator Award from the California Institute for Regenerative Medicine, a Glenn Foundation for Medical Research Award, an Ellison Medical Foundation Senior Scholar Award, and the 2012 Vincent Cristofalo 'Rising Star' Award in Aging Research. This fall, she was awarded a Pioneer Award from the NIH Director's fund, an award that supports scientists of exceptional creativity, who propose pioneering and transforming approaches to major challenges in biomedical research.

Andrei V. Budanov, PhD

A.V.B. is an Assistant Professor of the Molecular Biology of Cancer and Aging in the Department of Human and Molecular Genetics at Virginia Commonwealth University. He received his PhD in 2002 and was a postdoctoral fellow in the lab of Michael Karin at the University of California, San Diego, until 2011. He was a recipient of several fellowships and awards, such as postdoctoral fellowship from the Tobacco-Related Disease Research Program, K99/R00 NCI NIH career development grant, W.E. Lower prize for the best publication at the Cleveland Clinic Foundation, and Oxygen Club California BASF Young Investigator Award. He established his own laboratory in 2011.

Tao Cheng, MD

T.C. received his medical degrees and clinical training (internal medicine and hematology) from the Secondary Military Medical University, Shanghai, China. He did his postdoctoral fellowship (hematopoiesis and stem cell biology) at the Massachusetts General Hospital and Harvard Medical School with Dr. David Scadden. He is now a Professor of Stem Cell Biology and Regenerative Medicine at the Peking Union Medical College, deputy director of Institute of Hematology and Blood Diseases Hospital at the Chinese Academy of Medical Sciences and also director of the State Key Laboratory of Experimental Hematology in China. Prior to his current tenure, he was Assistant Professor of Medicine at Harvard Medical School and ranked up from Assistant Professor to Tenured Professor at University of Pittsburgh School of Medicine. He has received many awards, including the Young Faculty Scholar Award from the American Society of Hematology in 2002, the Chang-Jiang Scholarship from the Ministry of Education of China in 2007, the Scholar Award from the Leukemia and Lymphoma Society of the USA in 2008 and the Distinguished Professorship Award from the China Medical Board in 2008.

T.C.'s research mainly concerns cell cycle control of stem cells and stem cell response to injury or disease. His focus is on (1) the roles of cell cycle regulators (mainly CDK inhibitors) in stem cell self-renewal, (2) hematopoietic stem cell protection in transplant recipients or under pathological conditions, and (3) reprogramming potential of normal blood cells versus leukemia cells.

Gerald de Haan, PhD

G.d.H. is a Professor of Molecular Stem Cell Biology at the Department of Cell Biology, University Medical Center Groningen, The Netherlands (www.rug.nl/umcg), and co-director of the European Research Institute on the Biology of Aging (www.eriba.umcg.nl). He received his PhD in 1995 and was a postdoctoral fellow in the lab of Gary Van Zant at the University of Kentucky until 1998. He was awarded a fellowship by the Royal Netherlands Academy of Arts and Sciences to establish his own lab in Groningen and received a VICI grant from the Netherlands Organization for Scientific Research.

G.d.H.'s research interests relate to the molecular understanding of self-renewal of hematopoietic stem cells. This includes the identification of genes that regulate self-renewal, hematopoietic stem cell expansion, development of leukemia, and studies on hematopoietic stem cell aging. In earlier studies, G.d.H.'s group was able to show how hematopoietic stem cell turnover is correlated with mouse lifespan. Genetic studies have identified genomic loci that control these parameters, and more recently genome-wide expression studies have resulted in the construction of gene networks that underlie stem cell turnover and functioning. In addition, his lab studies the involvement of epigenetic modifications during hematopoietic stem cell aging.

Jan van Deursen, PhD

J.v.D. received his PhD in Cell Biology at the University of Nijmegen, The Netherlands, in 1993. He started his own lab at St. Jude's Children's Research Hospital in 1996. He joined Mayo Clinic in 1999, where he is currently a Professor of Biochemistry/Molecular Biology and Pediatrics at Mayo Clinic, Rochester, Minn., USA. He is the Vita Valley Named Professor of Cellular Senescence and directs the Senescence Program of the Robert and Arlene Kogod Center on Aging, the Cell Biology Program of the Mayo Clinic Comprehensive Cancer Center, and the Mayo Clinic Gene Knockout and Transgenic Core Facility. Beginning in early 2012, he is also the Chair of the Biochemistry/Molecular Biology Department.

Connie J. Eaves, PhD

C.J.E. is a Distinguished Scientist in the British Columbia Cancer Agency's Terry Fox Laboratory in Vancouver, B.C., Canada. She was a co-founder of the Terry Fox Laboratory and since has served both as its Deputy Director and Director. She is also a Professor in the Department of Medical Genetics at the University of British Columbia and has held many other senior leadership positions. These include election as President of the International Society for Experimental Hematology, and being appointed President of the former National Cancer Institute of Canada, Vice-President of Research at the BCCA, Associate Dean of Research in Medicine at the University of British Columbia, and member of the Board of Genome Canada. She originally trained in Biology, Chemistry and Genetics at Queens University in Canada and obtained doctoral and postdoctoral training in immunology, experimental hematology and medical biophysics in the UK and Canada. She has received numerous national and international awards, including receipt of the NCI

(Canada) Robert L. Noble Prize for Excellence in Cancer Research, the American Society of Hematology Henry Stratton Medal for Lifetime Achievement and the 2013 Rowley Prize of the International CML Foundation. These recognize her pioneering research in the development and use of robust, quantitative methods to detect, purify and characterize normal and malignant progenitors and stem cells of the blood and breast and their mechanisms of normal and perturbed control. Most recently, her group has led the discovery of heterogeneous stem cell states in these populations and identified the Lin28-Let7-HMGA2 axis as an important regulator of developmental changes in blood stem cell properties. She has also introduced the use of barcoding to analyze the clonal outputs of human cells with in vivo blood and breast tissue regenerative activity.

Bruce Edgar, PhD

B.E. was born in 1960 and grew up in California. He received a BA in biology from Swarthmore College 1982, and a PhD in genetics from the University of Washington in 1987, where he studied with Gerold Schubiger. B.E. did postdoctoral work at UCSF and Oxford University with Pat O'Farrell and Paul Nurse, respectively, and in 1993 became an independent investigator at the Fred Hutchinson Cancer Research Center in Seattle. In 2009, he moved to Heidelberg, Germany, where he is currently a Professor in the Center for Molecular Biology at the University and a division leader at the German Cancer Research Center. Since about 1986, Dr. Edgar's research has used the *Drosophila* model system to study how cell growth and proliferation are controlled during development, tissue maintenance, and tumorigenesis. These studies involve the genetic analysis of cell cycle control, growth-related metabolism, transcriptional programming, signal transduction, and cell-cell communication. A number of key cell cycle and growth regulatory genes have been identified and characterized, some of which play important roles in human disease, most notably cancer. Current studies in Dr. Edgar's lab focus on how the fly's intestinal stem cells support regenerative growth, and how these stem cells can become tumorigenic.

Hartmut Geiger, PhD

H.G. is Director of the Institute for Molecular Medicine at Ulm University, Ulm, Germany, as well as an Adjunct Full Professor in the Division of Experimental Hematology and Cancer Biology at the Cincinnati Children's Hospital Medical Center in Cincinnati, Ohio, USA.

After obtaining his PhD at the Max Planck Institute of Immunobiology in Freiburg, Germany, H.G. moved to the laboratory of Gary Van Zant at the University of Kentucky in Lexington, Ky., USA, to study the genetic regulation of hematopoietic stem cells. He subsequently was appointed to the Division of Experimental Hematology at Cincinnati Children's Hospital Medical Center where he currently holds adjunct status. He was appointed, after leading the KFO 142 in the Division of Dermatology and Allergic Diseases, to his current leadership position as the director of the Institute for Molecular Medicine in Ulm in 2013.

H.G.'s research focuses on hematopoietic stem cell biology, with a special emphasis on molecular pathways of stem cell aging and alterations of stem cell niche interaction upon aging.

Margaret A. Goodell, PhD

M.A.G is a Professor and director of the Stem Cells and Regenerative Medicine Center at Baylor College of Medicine in Houston, Tex., USA. M.A.G received her doctorate from Cambridge University, UK, and underwent postdoctoral training at Massachusetts Institute of Technology (MIT) and Harvard Medical School. At MIT, she developed the 'side population' or 'SP' method, commonly used for enrichment of hematopoietic stem cells. Her current research is focused on the fundamental mechanisms that regulate normal hematopoietic stem cells. Recent work has focused particularly on how interferon-γ and infectious stress stimulate stem cells, as well as epigenetic mechanisms of their regulation.

M.A.G has been on the faculty of the Baylor College of Medicine since 1997 as a member of the Center for Cell and Gene Therapy, and the Departments of Pediatrics, Molecular and Human Genetics, and Immunology. She holds the Vivian L. Smith Chair in Regenerative Medicine. She received the Stohlman Scholar Award from the Leukemia and Lymphoma Society in 2006, the O'Donnell Award from TAMEST in 2011, and the Damashek Prize from the American Society of Hematology in 2012. M.A.G has served on the board of the International Society for Stem Cell Research (2005–2008) and the International Society for Experimental Hematology (2009–2012). She is on the editorial boards of *Cell Stem Cell* and *PLoS Biology* and serves as an Associate Editor for *Blood*. She is President of the International Society of Hematology. M.A.G directs a laboratory of about 15 students and postdoctoral fellows.

Heinrich Jasper, PhD

H.J. is Professor at the Buck Institute for Research on Aging in Novato, Calif., USA. He obtained a diploma in biochemistry from the University of Tübingen, Germany, in 1999 and his PhD from the University of Heidelberg/EMBL, Germany, in 2002, where he studied transcriptional regulation of developmental processes in *Drosophila*. H.J. assumed his first faculty appointment (Research Assistant Professor) at the Department of Biomedical Genetics of the University of Rochester Medical Center in 2003, and in 2005 was appointed Assistant Professor at the Department of Biology of the School of Arts, Sciences, and Engineering of the University of Rochester. He moved to the Buck Institute in July 2012.

H.J.'s studies focus on the role of stress signaling in regulating various aspects of physiology that influence lifespan in *Drosophila*. His work has demonstrated that an antagonistic interaction between the stress-regulated Jun-N-terminal kinase pathway and insulin signaling regulates metabolic homeostasis, apoptosis, growth, and stress resistance in flies and significantly affects lifespan. Most recently, his lab has initiated studies to characterize the role of stem cells and proliferative homeostasis in maintaining the health and lifespan of flies, establishing newly identified regenerative processes in the *Drosophila* midgut as a model for understanding the importance of somatic stem cell function for the lifespan of metazoans.

H.J. received the Senior Fellow Award of the Ellison Medical Foundation in 2008 and was named Wilmot Assistant Professor in Arts, Sciences, and Engineering in 2009. He has published numerous research papers, and his work was and is funded by the American Federation for Aging Research, the National Institute of Aging, the National Eye Institute, the New York Stem Cell Initiative, and the Ellison Medical Foundation.

D. Leanne Jones, PhD

D.L.J. is in the Department of Molecular, Cell, and Developmental Biology at the University of California, Los Angeles. After completing her PhD in microbiology and molecular genetics at Harvard Medical School with Dr. Karl Münger, she engaged in postdoctoral studies with Philip Ingham at the Medical Research Council Centre for Developmental Genetics in Sheffield, UK, and then with Margaret Fuller in the Department of Developmental Biology at the Stanford University School of Medicine in Palo Alto, Calif., USA. She started her own research group at the Salk

Institute for Biological Studies in 2004, and she moved her group to UCLA in 2013. D.L.J. has received several research awards including ones from the Ellison Medical Foundation, the American Cancer Society, and the California Institute for Regenerative Medicine. D.L.J.'s research focuses on the molecular mechanisms underlying the manner in which aging affects stem cells, the stem cell environment (niche), and the relationship between the two. Using *Drosophila melanogaster* as a model system for studying the aging of adult stem cells, her lab found that significant changes to the stem cell niche occur, which are accompanied by a concomitant loss of stem cells. An aging-related decline in the expression of key stem cell self-renewal factors normally produced by supporting niche cells revealed that stem cell niches are highly dynamic, rather than static, structures. These studies suggest that genetic programs are in place to regulate maintenance of a functional stem cell niche over time. Therapeutic strategies that manipulate the size and activity of stem cell niches will complement stem cell transplantation in regenerative medicine and the treatment of cancer.

Jan Karlseder, PhD

J.K. is the Shiley Chair Professor at the Salk Institute for Biological Studies in La Jolla, Calif., USA. After completing his studies in molecular biology and biochemistry at the University of Vienna, Austria, he joined the laboratory of Titia de Lange at the Rockefeller University in New York for postdoctoral studies. In 2002, J.K. moved to the Salk Institute as Assistant Professor, was promoted to Associate Professor in 2007 and to Full Professor in 2012. He is currently the director of the Glenn Center for Research on Aging at the Salk, a member of the Dulbecco Cancer Center in the Molecular and Cellular Biology Department, and an Adjunct Professor at the Moores Cancer center at the University of California, San Diego.

J.K.'s laboratory focuses on various aspects of telomere function and the interaction of telomeres with the cell cycle and checkpoint machineries. The lab's main contributions during the last years were (1) that the Werner helicase is required for replication of the telomeric lagging strand and by that action preserves genome stability, (2) that the DNA damage machinery is recruited to functional telomeres after their replication to establish a protective structure at chromosome ends, (3) signaling from shortening telomeres during aging downregulates histone synthesis, thereby changing chromatin structure throughout the genome, (4) mitotic inhibition causes telomere deprotection, explaining why treatment with mitotic inhibitors induces a DNA damage response, (5) differential signaling between telomeres and DNA breaks allows cells to arrest with a diploid genome in G1 of the cell cycle, and (6) that alternative lengthening of telomeres is a function of inappropriate chromatin assembly at telomeres.

Michael Milyavsky, PhD

M.M. is a senior scientist at the Department of Pathology, Sackler Faculty of Medicine, Tel Aviv University, Israel. He received his PhD in 2005 from the Weizmann Institute of Science in Israel and was a postdoctoral fellow in the laboratory of Prof. John E. Dick at the Ontario Institute for Cancer Research, Toronto, Ont., Canada, until 2012. M.M. has received several research awards for his postdoctoral research, including long-term EMBO fellowship and European Hematology Association award. His main contribution includes the demonstration of distinct DNA damage response characteristics of human hematopoietic stem cells (HSCs).

M.M's independent research at Tel Aviv University focuses on the molecular mechanisms responsible for HSCs' responses to DNA damage. MM's laboratory pursues the following scientific questions: (1) how HSCs cope with DNA damage to preserve normal blood regeneration and to limit the risk of leukemogenesis; (2) how leukemia stem cells escape therapy, and (3) what are the strategies to eliminate residual leukemia cells to ensure leukemia cure? In 2012, he was awarded FP7 Marie Curie Career Integration Grant to study DNA damage responses in human HSCs.

Huck-Hui Ng, PhD

H.-H.N. is the Executive Director of the Genome Institute of Singapore. The Genome Institute of Singapore was started as a national flagship program to tap into the innovation and impact of deciphering the genetic blueprint of mankind. Today, it houses more than 200 researchers working on different aspects of human genomics (human genetics, infectious diseases, cancer therapeutics and stratified oncology, stem cell genomics, cancer stem cell genomics, translational genomics, computational and systems biology).

H.-H.N. graduated from the National University of Singapore with a First-Class Honors degree in Molecular and Cell Biology, and obtained his PhD from the University of Edinburgh. He spent the next few years working at the Harvard Medical School as a Damon Runyon-Walter Winchell research fellow. Upon return to Singapore, he started his research program on Stem Cell Genomics at the Genome Institute of Singapore.

His lab works on different aspects of Systems Biology of Stem Cells. Specifically, his group uses genome-wide approaches to dissect the transcriptional regulatory networks in embryonic stem cells with the aim to identify key nodes in this network. This had led to the first paper on the whole genome and unbiased mapping of key transcription factors in mouse embryonic stem cells. His group also conducted the whole genome genetic screen for human embryonic stem cells. More recently, his lab has begun to investigate the reprogramming code behind the induction of pluripotency in somatic cells. H.-H.N.'s works have been published in journals such as *Cell*, *Science* and *Nature*. He also sits on the editorial board of international journals such as *Genes & Development*. His papers have received over 12,000 citations and his H-index is 44.

His research work has earned him several prestigious national and international accolades including the Singapore Youth Award (2005 and 2010), the National Science Award 2007, the HUGO Chen's New Investigator Award 2010 and the President's Science Award 2011.

H.-H.N. holds Adjunct Professor appointment at the National University of Singapore (Departments of Biochemistry and Biological Sciences) and the Nanyang Technological University (School of Biological Sciences). He is also the President of the Stem Cell Society, Singapore.

Andre Nussenzweig, PhD

During the course of his career, A.N. has made major contributions to our understanding of how the integrity of the genome is maintained. A.N. has made a series of incisive discoveries, including: establishing that the nonhomologous end-joining pathway acts as a genomic 'caretaker' that protects against cancer [Nature 2000;404:510–514]; finding that the DNA damage response is activated under physiological processes of antigen receptor rearrangements [Science 2000;290: 1962–1965; Nature 2001;414:660–665]; determining the etiology of chromosomal translocations [Nature 2006;440:105–109; Cell 2008;135: 1028–1038]; finding that a core histone can act as a tumor suppressor [Science 2002;296:922–927; Cell 2003;5:675–679]; defining a new molecular branch of the DNA damage G2/M checkpoint [Nat Cell Biol 2002;4:993–997]; understanding the mechanism by which irradiation induces foci form [Nat Cell Biol 2002;4:993–997]; uncovering the first protein that is essential for the transcriptional inactivation of the XY chromosomes in male meiosis [Dev Cell 2003;4:497–508]; discovery of two complementary genome maintenance functions in DNA repair and apoptosis that prevent genetic damage from being passed on from one generation to the next [Cell 2007;130:63–75]; discovery of a methyltransferase complex that promotes chromatin accessibility critical for lymphocyte rearrangements [Science 2010;329: 917–923], and the discovery of cellular mechanisms that regulate the choice of repair pathways critical for preventing breast cancer [Cell 2010;41: 27–38].

Emmanuelle Passegué, PhD

E.P. is an Associate Professor of Medicine in the Hematology/Oncology Division in the Eli and Edythe Broad Center for Regenerative Medicine and Stem Cell Research at the University of California, San Francisco. E.P. received her PhD in Molecular and Cellular Endocrinology from the University Paris XI, France. She then trained as a mouse geneticist with Dr. Erwin Wagner (Institute for Molecular Pathology, Vienna, Austria) and as a stem cell biologist with Dr. Irv Weissman (Stanford University) before joining the UCSF faculty in 2006. Her research investigates the biology of blood-forming hematopoietic stem cells in normal and deregulated contexts such as development of hematological malignancies and physiological aging. Recent work from her laboratory investigates the function of autophagy in the metabolic control of hematopoietic stem cell activity. E.P. has received a number of awards, including a scholar award from the American Society of Hematology, a scholar award from the Rita Allen Foundation, a new faculty award from the California Institute for Regenerative Medicine, and a scholar award from the Lymphoma and Leukemia Society.

Pier Giuseppe Pelicci, MD, PhD

P.G.P. is scientific co-director of the European Institute of Oncology, chairman of the Department of Experimental Oncology at the European Institute of Oncology (IEO), Milan, Italy, and scientific director of the SEMM Foundation (European School of Molecular Medicine, Milan, Italy). IEO is a comprehensive cancer center focused on advanced treatments, diagnostics, clinical trials, cancer prevention, training, higher education and advanced research. SEMM is a private foundation whose mission is to foster innovative education in molecular medicine, medical nanotechnology and bio-ethics. At IEO, P.G.P. is responsible for the strategic planning of the IEO research programs, including basic, translational (Molecular Medicine Program) and clinical research. At SEMM, P.G.P. is responsible for the development of three PhD programs (Molecular Medicine, Medical Nanotechnology, Life Sciences: Foundations & Ethics).

P.G.P. is a member of the American Association for Cancer Research, the European Molecular Biology Organization, the European Haematology Association, the European Society for Engineering and Medicine, the European Cytokine Society, the New York Academy of Sciences, the American Society for Microbiology, the Italian Association of Biophysics and Molecular Biology, and the Italian Society of Cancerology. He is past President (1998–2000) of the Italian Society of Experimental Hematology.

P.G.P. was honored with several prestigious international fellowships and awards, such as the 'C. Cioffrese' Prize for Cancer Research (Fondazione Carlo Erba, Italy), the 'Foundation Chiara d'Onofrio' (Italy), the 'Guido Venosta' Prize of the Italian Foundation for Cancer Research, the Award for 'Excellence in Medicine' of the American-Italian Foundation for Cancer Research (New York, N.Y., USA), and the H.S. Raffaele Prize (Italy). He is presently Full Professor of Pathology at the University of Milan and co-founder of the Biotech holding Genextra. Genextra controls four Biotech companies (Congenia, DAC, Tethis and Intercept).

P.G.P. is co-founder and co-director of the IFOM-IEO Campus, a research infrastructure that hosts IFOM, the IEO laboratory research activities, SEMM, Genextra and Cogentech.

Thomas A. Rando, MD, PhD

T.A.R. is Professor of Neurology and Neurological Sciences at Stanford University School of Medicine (www.stanford.edu/~casco). He is director of the Glenn Laboratories for the Biology of Aging at Stanford, and he is director of the Center for Tissue Regeneration, Repair, and Restoration at the VA Palo Alto Health Care Systems, where he is also chief of the Neurology Service. He received his AB degree in biochemistry, an MD degree, and a PhD degree in Cell and Developmental Biology from Harvard University. He completed his clinical training in neurology at the University of California, San Francisco, and postdoctoral training in the Department of Molecular Pharmacology at Stanford University, where he was a Howard Hughes physician postdoctoral scholar. T.A.R. has received numerous awards and has been recognized by academic and professional societies for his work. He received a Paul Beeson Physician Faculty Scholar in Aging Award from the American Federation for Aging Research and an Ellison Medical Foundation Senior Scholar Award. In 2005, he received an NIH Director's Pioneer Award for his work at the interface between stem cell biology and the biology of aging.

T.A.R.'s lab focuses on the biology of skeletal muscle stem cells in adult muscle homeostasis, aging, and disease. In addition, his laboratory focuses on the pathogenesis and treatment of muscular dystrophies, with particular emphasis on cell and gene therapy. His laboratory described the critical roles of Notch and Wnt signaling in the lineage progression of adult muscle stem cells or satellite cells. Groundbreaking work from his lab showed that the age-related decline in stem cell function is due primarily to influences of the aged environment, rather than to intrinsic aging of the stem cells themselves. These findings have profound implications for the emerging field of regenerative medicine, both for enhancement of endogenous tissue repair and for stem cell transplantation approaches. His laboratory continues to focus on adult muscle stem cells, with recent interests in epigenetic regulation and on the role of microRNAs in quiescence and activation.

Hans-Reimer Rodewald, PhD

H.-R.R. carried out his PhD in immunology at the MPI for Immunobiology in Freiburg working on T cell receptor-biased T cell responses, and spent his postdoctoral years at the Dana-Farber Cancer Institute, Boston, Mass., USA, focusing on T cell receptor isoforms and early T and NK cell development. In 1992, he started his own laboratory as a member of the Basel Institute for Immunology, where he identified T cell- and mast cell-committed progenitors, and characterized growth factors (Kit) and cytokines that are crucial in lymphopoiesis, erythropoiesis, and myelopoiesis. H.-R.R. has been head of the Institute for Immunology at the University of Ulm from 1999 to 2010, and is now leading the Division of Cellular Immunology and coordinating the 'Tumor Immunology Program' at the German Cancer Research Center in Heidelberg. The Rodewald group is currently focusing on fate mapping of hematopoietic stem cells and lineages, on thymus biology, and origins of T cell leukemia, and on physiological and pathological functions of mast cells.

Derrick Rossi, PhD

D.R. received his BSc and MSc degrees from the University of Toronto, in the 1990s, and his PhD from the University of Helsinki, Finland, in 2003. He trained as a postdoctoral fellow at Stanford University in the laboratory of Dr. Irving Weissman from 2003 to 2007. D.R. is an Assistant Professor in the Stem Cell and Regenerative Biology Department at Harvard Medical School, and Harvard University. He is also an investigator in the Program in Cellular and Molecular Medicine at Boston Children's Hospital, and is a principal faculty member of the Harvard Stem Cell Institute.

The Rossi lab has a profound interest in understanding the mechanisms enabling self-renewal and multipotency in HSCs, which we study using cellular, molecular, genetic, small molecule, and epigenetic approaches. It is also interested in understanding the extent to which the aging of hematopoietic stem and progenitor cells contributes to the pathophysiological conditions arising in the aged hematopoietic system. The lab is also pursuing several lines of investigation aimed at reprogramming the cellular identity of a number of cell types into clinically useful cell types including HSCs through various approaches including the use of novel technologies developed in the lab.

D.R. has received multiple awards including the Robertson Investigator Award from the New York Stem Cell Foundation. The *Time* magazine cited D.R.'s discovery of modified-mRNA reprogramming as one of the top ten medical breakthroughs of 2010. The *Time* magazine also named D.R. as one of the *'People Who Mattered'* in 2010, and as one of the 100 Most Influential People *(Time 100)* in 2011. In 2010, D.R. co-founded Moderna Therapeutics, a Cambridge-based company developing modified-mRNA therapeutics. He is also a co-founder of CRISPR Therapeutics, a Switzerland-based company focused on developing Cas9/CRISPR-based therapeutics.

Michael Rudnicki, PhD

M.R. is a senior scientist and the director of the Regenerative Medicine Program and the Sprott Centre for Stem Cell Research at the Ottawa Hospital Research Institute. He is Professor in the Department of Medicine at the University of Ottawa. M.R. is the Scientific Director of the Canadian Stem Cell Network. He is a Fellow of the Royal Society of Canada, and holds the Canada Research Chair in Molecular Genetics. He is an Associate Editor of *Cell Stem Cell* and the *Journal of Cell Biology*, and is Co-Editor in Chief of *Skeletal Muscle*. M.R. has organized international research conferences as one of the founding directors of the Society for Muscle Biology.

M.R.'s laboratory works to understand the molecular mechanisms that regulate the determination, proliferation, and differentiation of stem cells during embryonic development and during tissue regeneration. The lab has conducted leading studies into both embryonic myogenesis and the function of muscle stem cells (satellite cells) in adult regenerative myogenesis. In particular, they have worked extensively to understand the molecular mechanisms that regulate the function of satellite cells in skeletal muscle. Towards this end, the lab employs molecular genetic and genomic approaches to determine the function and roles played by regulatory factors. They identified Pax7 as a transcription factor required for the specification of satellite cells, and identified Wnt7a signaling as playing an important role in muscle stem cell function. His research has been published in scientific journals that include *Cell*, *Nature Cell Biology*, *Cell Stem Cell*, *Genes & Development*, and *PLoS Biology*.

Karl Lenhard Rudolph, MD

K.L.R. is scientific director of the Leibniz Institute for Age Research, Fritz Lipmann Institute, in Jena (www.fli-leibniz.de) since 2012 and appointed Professor for Molecular Age Research at the Medical Faculty of the Friedrich Schiller University. After completing medical studies in Göttingen and a residency in internal medicine with Michael P. Manns in Hannover, he engaged in postdoctoral studies with Ronald A. DePinho at the Albert Einstein College in New York and at the Dana-Farber Cancer Institute in Boston, Mass., USA. He headed an Emmy Noether Research Group in Hannover from 2001 to 2006 and the Max Planck Research Department at Ulm University from 2007 to 2012. K.L.R. has received several research awards including the Gottfried-Wilhelm Leibniz Award of the DFG in 2009, the Wilhelm Vaillant Award in Molecular Medicine in 2011 and the Science award 'Society Needs Science' of the Stifterverband for the German Science System 2012.

K.L.R.'s research focuses on the molecular mechanisms underlying the aging process, especially the aging of stem cells. He has carried out numerous projects on the functional consequences of telomere shortening and DNA damage accumulation. His main contributions include the demonstration that: (1) telomere dysfunction can reduce organ maintenance, stress responses, and the lifespan of mice, (2) telomere dysfunction limits organ regeneration and is associated with the evolution of cirrhosis in humans with chronic liver disease, (3) telomere shortening leads to chromosomal instability and increased development of tumors, (4) the deletion of specific DNA damage checkpoints can improve stem cell function, organ maintenance, and the lifespan of telomere dysfunctional mice without increasing cancer formation, (5) telomere dysfunction induces environmental alterations limiting the function of stem cells, and (6) DNA damage limits self-renewal of adult stem cells by inducing stem cell differentiation involving tissue-specific transcription factors.

Björn Schumacher, PhD

Since 2013, B.S. has been Professor for Genome Stability in Ageing and Diseases at the University of Cologne. He received his PhD at the Max Planck Institute for Biochemistry in Munich and conducted his postdoctoral research as EMBO and Marie Curie fellow at the Erasmus Medical Centre in Rotterdam. Since his appointment at the Cluster of Excellence: Cellular Stress Responses in Ageing-Associated Diseases at the University of Cologne in 2009, he received the innovation prize of the State of Northrhine-Westphalia, the European Research Council starting grant, and coordinates the Marie Curie initial training network on chronic DNA damage in ageing (CodeAge) of the EU. B.S. is secretary general of the German Society for Ageing Research and serves on several editorial boards. His research interest focuses on the molecular mechanisms through which DNA damage causally contributes to cancer development and ageing-associated diseases. Employing both the genetic *Caenorhabditis elegans* system and mammalian disease models, B.S. has uncovered cell-autonomous and systemic responses to genome instability. His work established a link between persistent DNA damage and systemic adaptations to DNA damage accumulation with ageing. Most recently, B.S. demonstrated that an ancestral innate immune response to DNA damage leads to systemic activation of stress resistance mechanisms, thus allowing the organism to adjust its physiology to the presence of tissue-specific genome instability.

Manuel Serrano, PhD

M.S. is a researcher at the Spanish National Cancer Research Centre (CNIO), in Madrid, and Director of the Molecular Oncology Program of the CNIO. After completing his studies and PhD in Madrid, M.S. joined the laboratory of David

Beach at Cold Spring Harbor Laboratory, Cold Spring Harbor, N.Y., USA, as postdoctoral fellow from 1992 to 1996. During this time, M.S. made one of his most important contributions with the discovery of the tumor suppressor p16. M.S. established his research group, first at the National Center of Biotechnology, Madrid, and since 2003 at the CNIO. The main contributions of M.S.'s laboratory during these years are related to the concept of oncogene-induced senescence and the anti-aging activity of tumor suppressors. More recently, M.S.'s group has reported on the relevance of tumor suppressors in metabolic syndrome, and it has demonstrated the feasibility of embryonic reprogramming in live adult organisms.

Toshio Suda, MD

T.S. received his medical degree from the Yokohama City University School of Medicine, Japan. He then completed residency at the Kanagawa Children's Medical Center in Yokohama and a postdoctoral fellowship as a research associate at the Institute of Hematology, Jichi Medical School, Tochigi, Japan, and in the Department of Medicine, Medical University of South Carolina, Charleston, S.C., USA. After returning to Japan, he worked as an Assistant Professor and then Associate Professor at the Jichi Medical School for 7 years. T.S. was appointed Professor at the Institute of Molecular Embryology and Genetics Department of Cell Differentiation, Kumamoto University School of Medicine. After 10 years in Kumamoto, he now works as a Professor of Developmental Biology at the Sakaguchi Laboratory, School of Medicine, Keio University, Tokyo, Japan (www.coe-stemcell. keio.ac.jp).

T.S. was able to clarify the mode of action of the colony-stimulating factors by establishing a single-cell manipulation, and he developed a new purification method for hematopoietic stem cells, 'KSL cells', and progenitors. This is currently being used as a standard method for the purification of hematopoietic progenitors. He was also able to identify the endosteal niche for hematopoietic stem cells, and subsequently established the new field of oxidative stress and stem cell aging. The interaction of stem cells and niches is considered to be one of the hot topics in stem cell and cancer stem cell biology. The findings of the T.S. laboratory might be applied in stem cell transplantation and treatment of patients with bone marrow failure and leukemia.

Elly M. Tanaka, PhD

E.M.T. has been Full Professor of Animal Models of Regeneration at the Technical University Dresden, DFG Research Center for Regenerative Therapies, since 2008. She studied as an undergraduate at Harvard University, performing her thesis work with Lawrence S.B. Goldstein. Her doctoral studies were performed at University of California, San Francisco, with Marc W. Kirschner where she applied digital technologies to imaging microtubules in neuronal axons. She then travelled to London as a Muscular Dystrophy Society and Helen Haye Whitney Foundation postdoctoral fellow to start her studies on salamander limb regeneration with Jeremy P. Brockes. There, she established molecular techniques to study serum factors that induce cell cycle reentry of regenerative salamander cells. Upon taking up a junior group leader position at the Max Planck Institute of Molecular Cell Biology and Genetics, Dresden, in 1999, she initiated live imaging and molecular genetics approaches to studying spinal cord and limb regeneration.

Her honors include the Biofutures Award from the German Federal Ministry of Biotechnology and Research (2003), as well as the European Research Council Advanced Investigator Award (2011).

E.M.T. is recognized for her work rejuvenating the study of regeneration biology. By developing molecular genetics and imaging techniques in the salamander, *Ambystoma mexicanum*, she has defined the stem cells that undertake limb regeneration and their potency. She has also performed clonal studies of the neural stem cells responsible for spinal cord regeneration and has identified the molecular pathways triggered by injury that induce their self-renewal and their regenerative phenotype. More recently, she has begun applying this knowledge toward directing the morphogenesis and differentiation of mouse and human neural stem cells.

Andreas Trumpp, PhD

A.T. is Full Professor and head of the Department for Stem Cells and Cancer at the German Cancer Research Center in Heidelberg, Germany (www.dkfz.de/en/stammzellen-und-krebs). He is also the founding managing director of the new Heidelberg Institute for Stem Cell Technology and Experimental Medicine gGmbH. He studied biology at the University of Freiburg (Germany), receiving his diploma in 1989 and his PhD in 1992 at the European Molecular Biology Laboratories in Heidelberg, Germany. In 1994, he moved to the USA where he did postdoctoral research in the laboratories of J. Michael Bishop and Gail R. Martin at the University of California, San Francisco. He joined the Swiss Institute for Experimental Cancer Research in Epalinges/Lausanne as an associate scientist in 2000, where he became head of the Genetics and Stem Cell Laboratory. In 2005, he became professor (PATT) of molecular oncology and stem cell biology at the École Polytechnique Fédérale de Lausanne before moving to Heidelberg.

Maarten van Lohuizen, PhD

M.v.L. received his PhD in 1992 from the University of Amsterdam, where he studied cooperating oncogenes in murine lymphomagenesis in the lab of Dr. A. Berns at the Netherlands Cancer Institute. After staying on for one year as a postdoc, he joined the group of Prof. Dr. Ira Herskowitz, University of California, San Francisco, for his postdoctoral training. In 1995, he returned to the Netherlands Cancer Institute as an Assistant Professor in the Division of Molecular Carcinogenesis. After his tenure in 2000, he joined the Division of Molecular Genetics in 2001, of which he was appointed head of division in 2002 (http://research.nki.nl/lohuizenlab). In addition, in 2001 he became a part-time professor on the subject of regulation of cell cycle control and oncogenesis at the Utrecht University Medical School and was appointed as a member of the Centre for Biomedical Genetics in 2003. In 2007, he became part-time professor at the University of Amsterdam Medical School with the profile 'biology and epigenetic regulation of normal and cancer stem cells'.

His group has made important contributions on the functional analysis of epigenetic gene silencing mechanisms by Polycomb group protein complexes, which play crucial roles in controlling development and cell fate and – when deregulated – contribute to cancer formation. Recently, his group has also developed high-throughput genome-wide genetic screens in cell-based assays and in cancer-prone mice (in collaboration with Prof. Dr. A. Berns and Prof. Dr. A. Bradley of the Sanger Center, UK) to identify new genes that contribute to cancer and classify them in functional groups/signaling pathways. By using the DamID technique, his group has recently comprehensively mapped all Polycomb target genes in the *Drosophila* genome (in collaboration with the group of Bas van Steensel). In addition, his group has recently demonstrated a crucial role for Bmi1/Pc-G protein complexes in maintaining hematopoietic, neuronal, breast epithelial and embryonic stem cell fate, and has implicated the Sonic Hedgehog morphogen as a regulator of Bmi1 expression in neuronal precursor.

Unraveling the role of Bmi1/Pc-G in stem cell fate versus differentiation decisions and the consequence of this for cancer (stem) cell biology is currently a major focus of his group.

Amy J. Wagers, PhD

A.J.W. directs a research laboratory at the Joslin Diabetes Center (www.joslinresearch.org) and the Harvard University Department of Stem Cell and Regenerative Biology (http://www.scrb.harvard.edu/node/24). She completed her PhD in immunology and microbial pathogenesis at Northwestern University in 1999 and received postdoctoral training in stem cell biology in Dr. Irving Weissman's lab in the Department of Pathology at the Stanford University School of Medicine. She is currently the Forst Family Professor of Stem Cell and Regenerative Biology, a principal faculty member of the Harvard Stem Cell Institute, and a member of Harvard Medical School's Paul F. Glenn Laboratories for the Biological Mechanisms of Aging. A.J.W. is a recipient of an Early Career Award from the Howard Hughes Medical Institute and the Presidential Early Career Award for Scientists and Engineers. She focuses on defining the factors and mechanisms regulating the migration, expansion, and regenerative potential of blood-forming and muscle-forming stem cells. Her lab employs cell sorting and transplantation approaches to assess stem cell phenotype and function and to determine the molecular mechanisms that cause the decline in stem cell activity seen typically in old age.

Rudolph KL (ed): Adult Stem Cells in Aging, Diseases and Cancer.
Else Kröner-Fresenius Symp. Basel, Karger, 2015, vol 5, pp 18–24 (DOI: 10.1159/000366565)

Stem Cells in Adult Intestine

Juan Feng · Tobias Sperka

Leibniz Institute for Age Research, Fritz Lipmann Institute, Jena, Germany

Abstract

The epithelium of the small intestine is organized into large numbers of self-renewing crypt-villus units. Intestinal stem cells reside in the crypts of Lieberkühn. In this chapter, we review current knowledge on intestinal stem cells. In addition, we summarize reports from the Else Kröner-Fresenius Symposium on advances of our understanding of stem cell function in the *Drosophila* intestine.

Intestine

The epithelium of the small intestine is organized into large numbers of self-renewing crypt-villus units. Six differentiated epithelial cell types are distinguished. The absorptive enterocyte (EC) is the most populous cell in the villus epithelium: it is a polarized columnar cell characterized by a luminal brush border. Goblet cells and enteroendocrine cells (EEs) secrete mucus and hormones, respectively, and are located both on villi and in crypts. Tuft cells also occur along the crypt-villus axis and may serve to sense luminal contents. Microfold cells are located in the specialized epithelium that overlies the Peyer's patches, accumulations of lymphoid cells that participate in mucosal immunity. Paneth cells are located at the base of the crypt of Lieberkühn and secrete bactericidal products such as lysozyme and defensins. Finally, epithelial intestinal stem cells (ISCs) reside in the crypts of Lieberkühn, give rise to all the aforementioned differentiated cell types, and thus drive self-renewal of the whole epithelium (fig. 1) [1].

Intestinal Stem Cells

In the intestine, each crypt contains a small number of ISCs; however, the molecular mechanisms that control their numbers within the ISC niche are incompletely understood. Models based on asymmetric stem cell division predominated the field; however, Snippert et al. [2] discovered that ISCs in the mouse intestine actually divide symmetrically, and that their decision to differentiate is not coupled to division. Some ISC divisions are duplicative (generating two ISCs), while others are doubly terminal (generating two transient cells committed to differentiate), and still others are asymmetric (generating one ISC and one committed cell). At the population level, the fre-

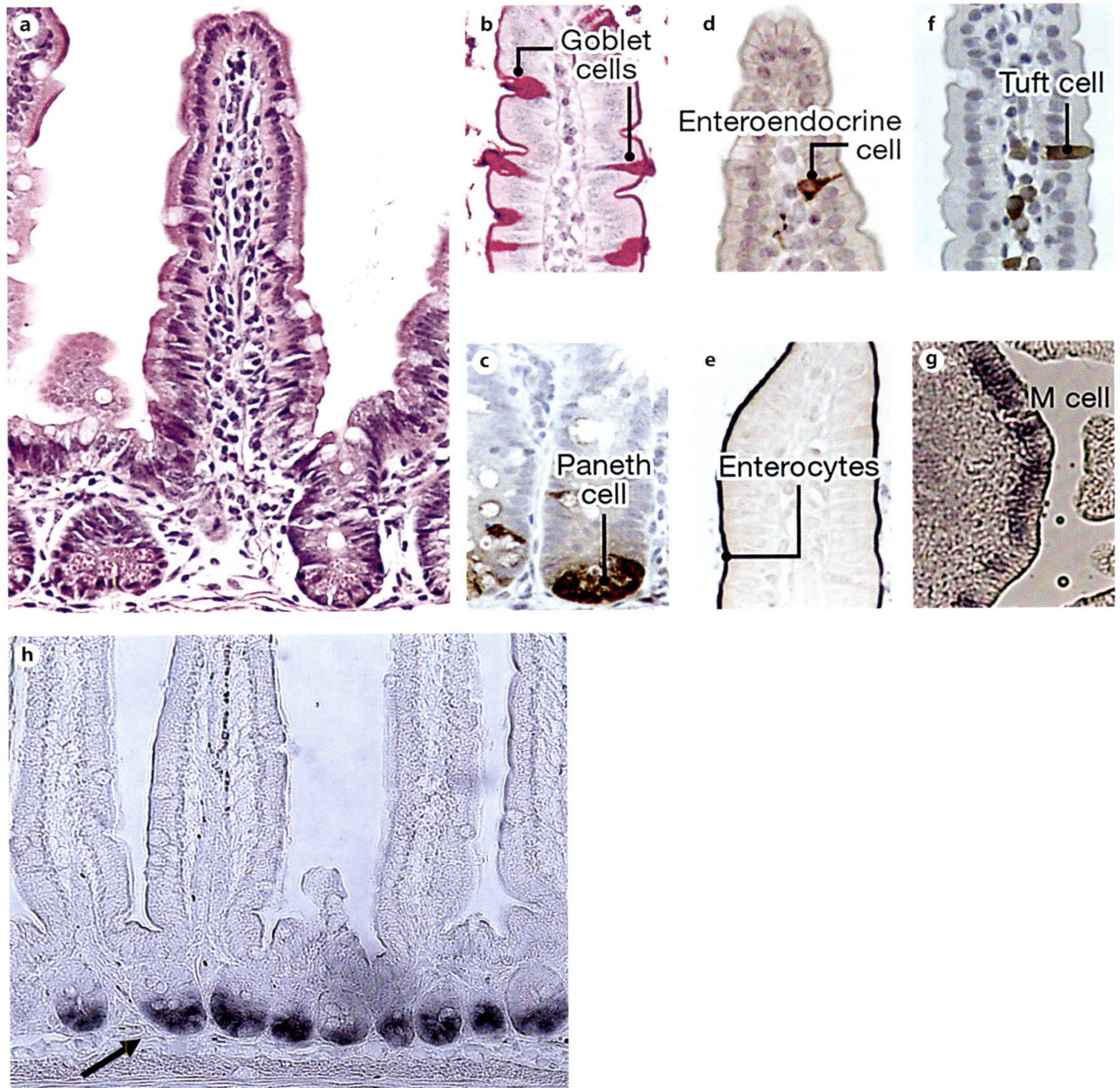

Fig. 1. Epithelial cell types of the small intestine. Different cell types of the mammalian small intestine are visualized by specific marker stainings. **a** Hematoxylin and eosin staining of the intestinal epithelium. **b** Periodic acid-Schiff-stained (purple) goblet cells on villus. **c** Lysozyme (brown)-stained Paneth cells at crypt bottoms. **d** Chromogranin-stained (brown) EE. **e** Alkaline phosphatase-stained (blue, at luminal brush borders) villus ECs. **f** DCAMKL1-stained tuft cell (courtesy of P. Jay). **g** Spi-B expression in microfold (M) cells. **h** Olfm4 in situ staining of stem cells in basal intestinal crypts (Juan Feng). From van der Flier and Clevers [1]. Reprinted with permission from *Annual Review of Physiology*.

quencies of duplicative and terminal divisions were found to be equal, explaining how a constant stem cell pool could be maintained.

To explain the balance of ISC loss and duplication, it was postulated that stem cells in the crypt base undergo 'neutral' competition for niche signals provided by a limited number of Paneth cells, which thereby define the number of stem cells each crypt can support [3]. Paneth cells are long-lived secretory cells that reside in the crypt base, intermingled with ISCs, and express regulators of ISC growth and survival, including EGF, TGF-α, Wnt3, and Dll4, thus serving as supportive niche cells for the ISCs. The mechanisms that determine the number of ISCs or Paneth cells are currently not known.

Secreted factors of the Wnt (wingless-related MMTV integration site) family play an important role in crypt proliferation [4]. In the absence of Wnt, free cytoplasmic β-catenin displays a short half-life due to the action of the APC (adenomatous polyposis coli) destruction complex. When Wnt proteins occupy their Frizzled-Lrp5/6 receptors, β-catenin is stabilized, accumulates, and translocates to the nucleus. It then engages Tcf transcription factors to activate transcription of Wnt/Tcf target genes. Tcf4 knockout mice lack proliferative crypts, indicating an important role for the Wnt/Tcf4 cascade in establishing and/or maintaining the epithelial stem cell compartment in the developing intestine [5]. Furthermore, transgenic expression of the Wnt receptor antagonist Dkk1 [6] or conditional deletion of β-catenin [7] or Tcf4 [8] showed that maintenance of adult crypt stem and progenitor cell proliferation continues to be dependent on Wnt/Tcf4 signaling. In addition, Wnt target genes Cyclin D1 [9] and cMyc [10] proved to be important for proliferation of undifferentiated cells in the crypt epithelium.

Notch signaling has turned out as another essential pathway instructing the intestinal epithelium. Several Notch receptors and their cognate ligands influence cell fate decisions. Inhibition of Notch receptor signal relay drives differentiation into a secretory fate causing a goblet cell metaplasia [11]. At the crypt base, Delta ligand-expressing Paneth cells ($Dll1^+Dll4^+$) trigger Notch1 and Notch2 on stem cells, thus repressing the transcription factor Math1. This inhibitory action restrains the Lgr5+ stem cells from terminal differentiation into the secretory lineage (fig. 2) [12–14].

Stem Cells in *Drosophila* Midgut

The discovery of somatic stem cells in adult *Drosophila*, particularly the ISCs of the midgut, has established *Drosophila* as an exciting model to study stem cell-mediated adult tissue homeostasis and regeneration. The endodermal portion of the intestinal epithelium in *Drosophila*, termed the midgut, is maintained by long-lived ISCs [15]. Compared with the mouse intestine, the fly midgut lacks crypts and villi but is composed of a cellular monolayer ensheathed by two orthogonal layers of visceral muscle. ISCs are accompanied by neighboring ECs and reside at the basal side of the epithelium facing a basement membrane, which is produced in part by visceral muscle [16, 17]. The ISC divides to self-renew and to give rise to committed progenitors (enteroblasts, EBs), which directly differentiate, without further amplifying cell divisions, into two functional cell lineages similar to those found in vertebrates: absorptive ECs and EEs. Differentiating ECs endoreplicate their genome 2–3 times to increase their size and develop a brush border similar to that in mammals. *Drosophila* lacks dedicated Paneth and Goblet cells, but some of their functions – such as immunity and barrier production – are fulfilled by ECs [16, 17].

The midgut ISCs adhere via integrin adhesion receptors to the visceral muscle [18]. The muscle in turn produces several factors that are capable of promoting ISC growth and maintenance, including wingless (a Wnt family member), vein (an EGF receptor ligand), and dilp3 (an insulin-

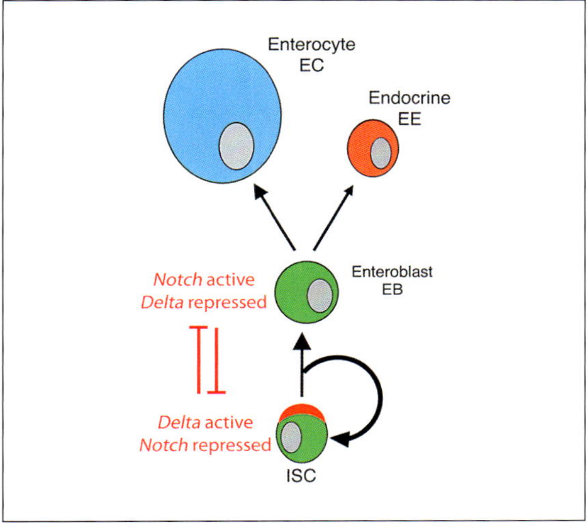

Fig. 2. Delta/Notch signaling in *Drosophila* midgut. The Notch pathway is pivotal for ISC homeostasis in the *Drosophila* midgut. Stem cells are the only Delta ligand-expressing cells and induce Notch signaling in the adjacent EB. The level of Delta expression furthermore instructs lineage choice to either EC or EE [28]. Inhibition of Notch blocks differentiation and causes a stem cell hyperplasia, whereas increased Notch activity supports differentiation and subsequent stem cell ablation [15, 28].

like peptide) [19]. Hence, the visceral muscle has been proposed to serve as the stem cell niche (fig. 3) [20, 21]. However, cells within the epithelial ISC lineage – differentiated ECs and transient EBs – also produce survival and growth factors that support *Drosophila's* ISCs. These factors are especially important during gut epithelial regeneration following damage, but they are also likely to be used for stem cell maintenance. During the development of the midgut, stem-like progenitor cells proliferate to form clusters, and some cells within these clusters differentiate into peripheral cells that function as a transient niche for the progenitors that build the adult gut and contribute the adult ISCs [22, 23]. These peripheral cells produce decapentaplegic, a BMP-type signal that suppresses differentiation [22], and the EGFR ligands Spitz and Keren [23], potent ISC growth factors. Thus, during midgut development, stem-like progenitor cells also generate an essential part of their own niche. The combined action of EGF receptor ligands (Spitz, Keren, Vein) and cytokines of the unpaired family (Upd) influences growth, survival, and proliferation of *Drosophila* ISCs and allows the epithelium to dynamically respond to environmental alterations, e.g. ranging from normal homeostasis over bacterial infections or intoxications [20, 21, 24–26]. On the other hand, ectopic activation of unpaired-induced Jak/Stat or EGFR signaling promotes uncontrolled stem cell proliferation, leading to severe midgut hyperplasia, whereas midguts defective in either pathway suffer reduced rates of epithelial renewal, leading in some cases to atrophy [27].

Notch and Delta

Notch is a central regulator in the *Drosophila* intestine. In the *Drosophila* midgut, Notch is expressed in progenitor cells, including ISCs and their undifferentiated daughter cell, the EBs. However, the Notch ligand Delta is expressed only in stem cells resulting in asymmetric Notch activation only in EBs [28]. Loss of Notch, Delta, or other pathway components such as Su(H) or Neuralized, or treatment with the γ-secretase inhibitor DAPT, all result in defective lineage commitment and a rapid, exponential expansion of stem-like cells that produce endocrine cells, but not ECs [29, 30]. Ectopic expression of activated Notch, in turn, drives the rapid, direct differentiation of ISCs into EC-like cells and consequently depletes the gut of stem cells [15].

This year's Else Kröner-Fresenius Symposium on molecular mechanisms of adult stem cell aging hosted two exiting presentations on *Drosophila* stem cell biology presented by Prof. Bruce Edgar from Heidelberg (DKFZ-ZMBH Alliance, Heidelberg, Germany) and Prof. Henri Jasper from Novato (Buck Institute for Research on Ag-

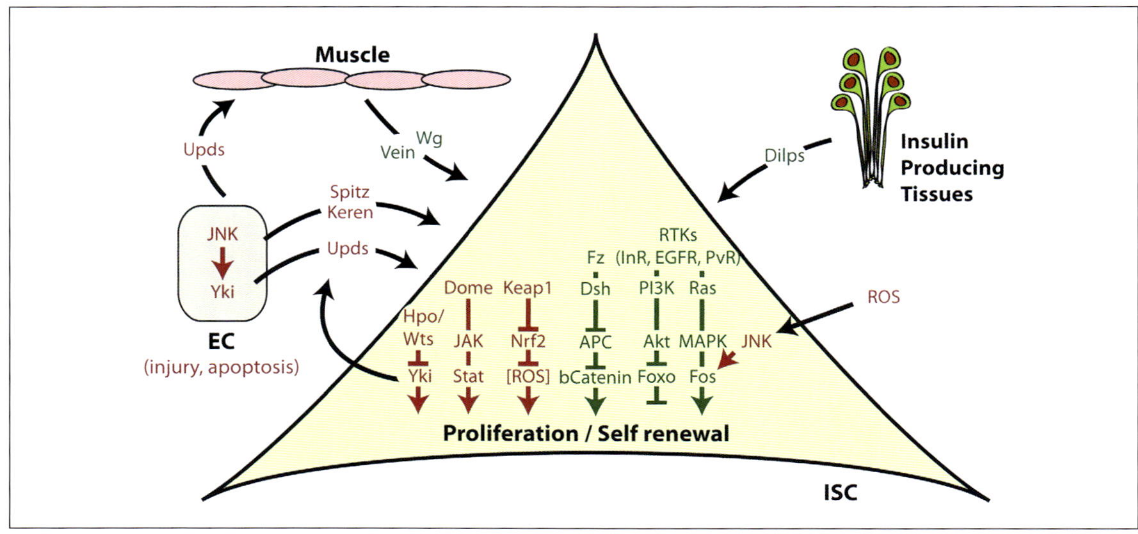

Fig. 3. Signaling pathways regulating ISC proliferation and self-renewal in *Drosophila*. ISCs integrate local and systemic cues with cell-intrinsic signals to adapt their proliferation rate to tissue demand. Signaling pathways required for homeostatic proliferation are represented in green, and pathways required for stress- and injury-induced ISC proliferation are represented in red. Reprinted from Biteau et al. [19] with permission from Elsevier.

ing, Novato, Calif., USA). Dr. Edgar presented data on signaling pathways in the *Drosophila* midgut mediating regeneration and homeostasis. Midgut ECs subjected to apoptosis, enteric infection, or JNK-mediated stress signaling, react by producing secreted cytokines of the Unpaired family (Upd, Upd2, and Upd3) (fig. 3) [26]. These ligands activate Jak/Stat signaling in ISCs, promoting their rapid division, thus enabling communication between the differentiated epithelium and the stem cell fraction [26]. Upd/Jak/Stat activity also promotes progenitor cell differentiation, in part by stimulating Delta/Notch signaling, and is required for differentiation in both unstressed and regenerating midgut. Hence, cytokine-mediated feedback enables stem cells to replace spent progenitor cells when they are lost, thereby establishing gut homeostasis (fig. 3) [26].

Dr. Edgar continued reporting remarkable work on stem cell tumors in the *Drosophila* intestine. Tumors form when Notch expression is repressed in ISCs, which in turn expand surprisingly rapidly due to elevated expression of the EGFR ligand Spitz, thus stimulating their own growth. Interestingly, Notch$^-$ stem cell tumors affect surrounding tissue dramatically utilizing signaling circuits, which are normally employed to control homeostasis. Expanding Notch$^-$ stem cell tumors displace ECs from the epithelium, which in turn activates JNK-mediated stress signaling in surrounding epithelium. In analogy to bacterial infection, JNK induces expression of Upd2, Upd3 cytokines in ECs, further stimulating tumor growth in a paracrine pattern. Additionally, Upd cytokines activate Vein (EGF) and dILP3 (IGF) in visceral muscle, which boosts tumor growth in a 2nd paracrine wave. The cytokine cocktail produced in response to Notch repression in ISCs is furthermore able to activate wild-type ISCs outside the tumor, thus driving a vicious cycle of uncontrolled proliferation. It will be interesting to follow up on these studies testing for relevance of feedback interactions between mutant stem cells and differentiated progeny in human pathology.

Other Signaling Pathways

The former work of Dr. Jasper's group showed age-associated changes of the *Drosophila* midgut. The studies suggest that the production of ROS in response to an increased bacterial population is the reason for the elevated JNK activity in the aged fly gut [31]. The consequences are elevated proliferation and imbalanced differentiation due to improper Notch/Delta signaling, underlining the detrimental effect of deficient lineage specification on tissue homeostasis. Their previous work elegantly showed that introduction of genetic mutations that moderately impair stem cell proliferation counterbalance stress-induced increases in JNK activity leading to the aging-associated breakdown in the maintenance of the intestinal epithelium [31]. This work actually stands in agreement with a new concept indicating that the rate of stem cell proliferation is a critical determinant for the lifespan of functional stem cells limiting organ homeostasis during aging.

During the Symposium, Dr. Jasper reported on Notch signaling controlling ISC differentiation in the *Drosophila* midgut, while ISC proliferation is regulated by growth factor signaling pathways, including insulin/IGF signaling (IIS) [32]. In the study, they explored the interaction between growth factor signals and Notch signaling in the control of ISC proliferation and differentiation. The work indicates that high expression of the TOR inhibitor TSC2 in the gut epithelium counterbalances TOR activation by the IIS pathway and helps to maintain growth-promoting TOR signaling in an inactive state. TSC2 expression thus has a critical role in sheltering ISCs from nutritional cues, ensuring their long-term maintenance. However, the Notch pathway is activated in EBs (progenitor cells generated as direct daughter cells from the ISCs) resulting in TSC2 repression and TOR de-repression. The experiments revealed that Notch-dependent repression of TSC2 is required and sufficient for differentiation of EBs into ECs, the absorptive cells of the epithelium. The findings establish a critical role for TSC in ISC maintenance and provide a mechanism by which Notch promotes differentiation into the EC fate. The human homologue of TSC2 is an important tumor suppressor, and the study provides new insight into how its regulation controls regenerative processes [33].

Outlook

The studies on the signaling pathways in ISCs will improve our understanding about stem cell proliferation, differentiation, self-renewal, and ageing, furthermore about adult tissue maintenance and regeneration, which will be helpful for regenerative medicine and therapeutic interventions of stem cell-derived tumors. The information from the studies of *Drosophila* ISC will be helpful to test these concepts in human ISC and analyze their contribution to ISC aging.

References

1 van der Flier LG, Clevers H: Stem cells, self-renewal, and differentiation in the intestinal epithelium. Annu Rev Physiol 2009;71:241–260.

2 Snippert HJ, van der Flier LG, Sato T, van Es JH, van den Born M, Kroon-Veenboer C, Barker N, Klein AM, van Rheenen J, Simons BD, Clevers H: Intestinal crypt homeostasis results from neutral competition between symmetrically dividing Lgr5 stem cells. Cell 2010;143:134–144.

3 Sato T, van Es JH, Snippert HJ, Stange DE, Vries RG, van den Born M, Barker N, Shroyer NF, van de Wetering M, Clevers H: Paneth cells constitute the niche for Lgr5 stem cells in intestinal crypts. Nature 2011;469:415–418.

4 Clevers H, Nusse R: Wnt/β-catenin signalling and disease. Cell 2012;149: 1192–1205.
5 Korinek V, Barker N, Moerer P, van Donselaar E, Huls G, Peters PJ, Clevers H: Depletion of epithelial stem-cell compartments in the small intestine of mice lacking Tcf-4. Nat Genet 1998;19: 379–383.
6 Pinto D, Gregorieff A, Begthel H, Clevers H: Canonical Wnt signals are essential for homeostasis of the intestinal epithelium. Genes Dev 2003;17:1709–1713.
7 Fevr T, Robine S, Louvard D, Huelsken J: Wnt/beta-catenin is essential for intestinal homeostasis and maintenance of intestinal stem cells. Mol Cell Biol 2007;27:7551–7559.
8 van Es JH, Haegebarth A, Kujala P, Itzkovitz S, Koo BK, Boj SF, Korving J, van den Born M, van Oudenaarden A, Robine S, Clevers H: A critical role for the Wnt effector Tcf4 in adult intestinal homeostatic self-renewal. Mol Cell Biol 2012;32:1918–1927.
9 Tetsu O, McCormick F: Beta-catenin regulates expression of cyclin D1 in colon carcinoma cells. Nature 1999; 398:422–426.
10 He TC, Sparks AB, Rago C, Hermeking H, Zawel L, da Costa LT, Morin PJ, Vogelstein B, Kinzler KW: Identification of c-MYC as a target of the APC pathway. Science 1998;281:1509–1512.
11 Milano J, McKay J, Dagenais C, Foster-Brown L, Pognan F, Gadient R, Jacobs RT, Zacco A, Greenberg B, Ciaccio PJ: Modulation of notch processing by gamma-secretase inhibitors causes intestinal goblet cell metaplasia and induction of genes known to specify gut secretory lineage differentiation. Toxicol Sci 2004;82:341–358.
12 Riccio O, van Gijn ME, Bezdek AC, Pellegrinet L, van Es JH, Zimber-Strobl U, Strobl LJ, Honjo T, Clevers H, Radtke F: Loss of intestinal crypt progenitor cells owing to inactivation of both Notch1 and Notch2 is accompanied by derepression of CDK inhibitors p27Kip1 and p57Kip2. EMBO Rep 2008;9:377–383.
13 Pellegrinet L, Rodilla V, Liu Z, Chen S, Koch U, Espinosa L, Kaestner KH, Kopan R, Lewis J, Radtke F: Dll1- and dll4-mediated notch signaling are required for homeostasis of intestinal stem cells. Gastroenterology 2011;140: 1230–1240.
14 Shroyer NF, Helmrath MA, Wang VY, Antalffy B, Henning SJ, Zoghbi HY: Intestine-specific ablation of mouse atonal homolog 1 (Math1) reveals a role in cellular homeostasis. Gastroenterology 2007;132:2478–2488.
15 Micchelli CA, Perrimon N: Evidence that stem cells reside in the adult *Drosophila* midgut epithelium. Nature 2006;439:475–479.
16 Casali A, Batlle E: Intestinal stem cells in mammals and *Drosophila*. Cell Stem Cell 2009;4:124–127.
17 Wang P, Hou SX: Regulation of intestinal stem cells in mammals and *Drosophila*. J Cell Physiol 2009;222:33–37.
18 Lin G, Zhang X, Ren J, Pang Z, Wang C, Xu N, Xi R: Integrin signaling is required for maintenance and proliferation of intestinal stem cells in *Drosophila*. Dev Biol 2013;337:177–187.
19 Biteau B, Hochmuth CE, Jasper H: Maintaining tissue homeostasis: dynamic control of somatic stem cell activity. Cell Stem Cell 2011;9:402–411.
20 Lin G, Xu N, Xi R: Paracrine Wingless signaling controls self-renewal of *Drosophila* intestinal stem cells. Nature 2008;455:1119–1123.
21 Xu N, Wang SQ, Tan D, Gao Y, Lin G, Xi R: EGFR, Wingless and JAK/STAT signaling cooperatively maintain *Drosophila* intestinal stem cells. Dev Biol 2011;354:31–43.
22 Mathur D, Bost A, Driver I, Ohlstein B: A transient niche regulates the specification of *Drosophila* intestinal stem cells. Science 2010;327:210–213.
23 Jiang H, Edgar BA: EGFR signaling regulates the proliferation of *Drosophila* adult midgut progenitors. Development 2009;136:483–493.
24 Biteau B, Jasper H: EGF signaling regulates the proliferation of intestinal stem cells in *Drosophila*. Development 2011; 138:1045–1055.
25 Jiang H, Grenley MO, Bravo MJ, Blumhagen RZ, Edgar BA: EGFR/Ras/MAPK signaling mediates adult midgut epithelial homeostasis and regeneration in *Drosophila*. Cell Stem Cell 2011;8:84–95.
26 Jiang H, Patel PH, Kohlmaier A, Grenley MO, McEwen DG, Edgar BA: Cytokine/Jak/Stat signaling mediates regeneration and homeostasis in the *Drosophila* midgut. Cell 2009;137:1343–1355.
27 Lin G, Xu N, Xi R: Paracrine unpaired signaling through the Jak/Stat pathway controls self-renewal and lineage differentiation of *Drosophila* intestinal stem cells. J Mol Cell Biol 2009;2:37–49.
28 Ohlstein B, Spradling A: Multipotent *Drosophila* intestinal stem cells specify daughter cell fates by differential notch signaling. Science 2007;315:988–992.
29 Perdigoto CN, Schweisguth F, Bardin AJ: Distinct levels of notch activity for commitment and terminal differentiation of stem cells in the adult fly intestine. Development 2011;138:4585–4595.
30 Bardin AJ, Perdigoto CN, Southall TD, Brand AH, Schweisguth F: Transcriptional control of stem cell maintenance in the *Drosophila* intestine. Development 2010;137:705–714.
31 Biteau B, Hochmuth CE, Jasper H: JNK activity in somatic stem cells causes loss of tissue homeostasis in the aging *Drosophila* gut. Cell Stem Cell 2008;3: 442–455.
32 Biteau B, Karpac J, Supoyo S, DeGennaro M, Lehmann R, Jasper H: Lifespan extension by preserving proliferative homeostasis in *Drosophila*. PLoS Genet 2010;6:e1001159.
33 Kapuria S, Karpac J, Biteau B, Hwangbo D, Jasper H: Notch-mediated suppression of TSC2 expression regulates cell differentiation in the *Drosophila* intestinal stem cell lineage. PLoS Genet 2012;8:e1003045.

Tobias Sperka, PhD
Leibniz Institute for Age Research, Fritz Lipmann Institute e.V. (FLI)
Beutenbergstrasse 11
DE–07745 Jena (Germany)
E-Mail tsperka@fli-leibniz.de

Rudolph K₋ (ed): Adult Stem Cells in Aging, Diseases and Cancer.
Else Kröne-Fresenius Symp. Basel, Karger, 2015, vol 5, pp 25–39 (DOI: 10.1159/000366566)

Molecular Mechanisms of Muscle Stem Cell Aging

Simon Schwörer • Stefan Tümpel

Leibniz Institute for Age Research, Fritz Lipmann Institute, Jena, Germany

Abstract

Muscle stem cells, resident beneath the basal lamina of skeletal muscle myofibers, are quiescent precursor cells and are required for normal postnatal growth and repair of skeletal muscles in adulthood. There is clear evidence of functional decline in skeletal muscle tissue during aging. This is accompanied by an impairment in muscle stem cell function and in the regenerative capacity of skeletal muscle during aging. The molecular mechanisms responsible for the impaired functionality of muscle stem cells during aging are not fully elucidated, but are likely to involve both cell-intrinsic changes as well as extrinsic alterations of the stem cell environment. Several molecular pathways have been implicated in the process of muscle stem cell aging. Here, we summarize recent progress in the modulation of the key molecular pathways and epigenetic modifications in both skeletal and cardiac muscle during aging.

Morphological and Molecular Properties of Muscle Stem Cells

Adult skeletal muscle consists predominantly of myofibers that are sustained by hundreds of myonuclei. During postnatal development, muscle stem cells divide to provide new myonuclei to the growing muscle fibers. Muscle-specific stem cells are also called satellite cells since they appear to orbit the outer surface of the muscle fiber (fig. 1) [1, 2]. They are defined by this unique anatomical location, underneath the basal lamina of mature, differentiated muscle fibers as mitotically quiescent cells [2]. They are morphologically characterized by their increased nuclear to cytoplasmic ratio, reduced organelle content and high amounts of heterochromatin, reflecting the mitotic and metabolic inactive state of these cells [3]. They are activated by injury or muscle growth and then become highly proliferative intermediate progenitor cells (fig. 1). During this process, the cell morphology changes; the cytoplasm extends, the amount of heterochromatin decreases and organelles become visible [4, 5]. Satellite cells generate mononucleated myoblasts that are capable of fusion and differentiation. Either the myoblasts align and fuse to form multinucleated myotubes to replace the injured muscle tissue, or they fuse with existing myofibers to repair damaged muscles [6–8].

Stem Cell Nature of Satellite Cells

The first hint of 'stemness' of satellite cells was given by Moss and Leblond [10], who showed that satellite cells are able to self-renew by undergoing asymmetric cell divisions, and thereby giving

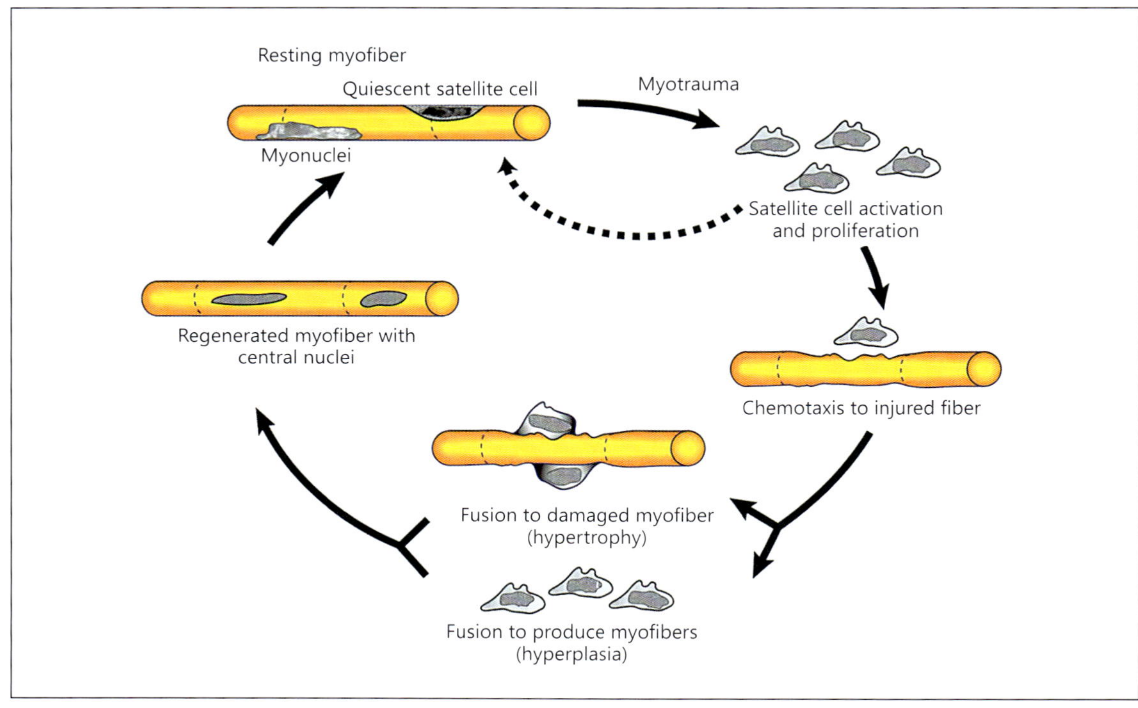

Fig. 1. Schematic diagram of satellite cell dynamics. The quiescent satellite cells are located beneath the basal lamina and are activated by injury. The daughter cells migrate from beneath the basal lamina and their progeny proliferates. They either fuse with the damaged myofibers or fuse with each other to produce new myofibers, while others self-renew to replenish the pool of quiescent satellite cells [9].

rise to both myonuclei and satellite cells. It was decades later that Collins et al. [7] demonstrated that satellite cells fulfill the criteria allocated to stem cells, the ability to self-renew and to form a differentiated cell type. Transplantation of a single muscle fiber with only a small number of associated satellite cells gave rise to hundreds of new muscle fibers with thousands of myonuclei. Moreover, the progeny of the grafted fiber repopulated its own stem cell pool with functional myogenic cells that could be reactivated in subsequent rounds of injury, thereby further contributing to tissue repair. The definitive proof of the self-renewal capacities of satellite cells was provided by the work of the Blau lab, showing that even a single satellite cell was able to self-renew and differentiate into proliferating progenitors after transplantation [11]. Eventually, serial transplantation studies which are routinely used to assess the long-term regenerative potential of hematopoietic stem cells, have been accomplished recently also for satellite cells [12]. In this study, it has been demonstrated that satellite cells were capable of seven rounds of serial transplantation, thereby giving evidence that they are indeed long-lasting stem cells.

Heterogeneity of Satellite Cells

Several studies have shown variability in the expression of phenotypic markers among satellite cells. One example is the heterogenic expression of Pax3 by satellite cells in different muscles of mice. In most hindlimb muscles, some forelimb and trunk muscles, satellite cells are negative for Pax3,

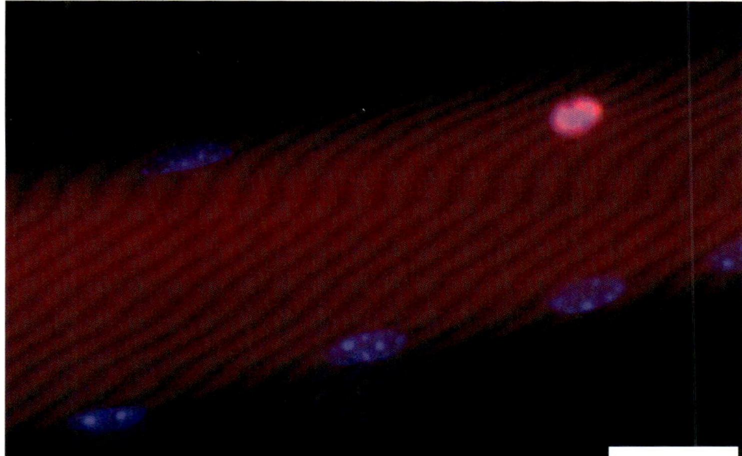

Fig. 2. Expression of Pax7 defines satellite cells within the myofiber. The myofiber was stained for Pax7 (red) and DAPI (blue). Bar = 20 μm.

whereas all satellite cells express Pax7 (fig. 2). Only a subset of satellite cells in muscles like those in the diaphragm, ventral trunk and parts of the forelimb is positive for Pax3 [13, 14]. Furthermore, it has been demonstrated that even Pax7 levels vary within the satellite cell pool [12]. Using a transgenic Pax7-GFP mouse, Rocheteau et al. [12] showed that there are satellite cells which display high Pax7-GFP levels (Pax7-GFP^{high}), while other satellite cells show low Pax7-GFP expression (Pax7-GFP^{low}). Pax7-GFP^{high} cells are less prone to myogenic differentiation compared to Pax7-GFP^{low} cells, and they can give rise to the latter after serial transplantations. In addition, Pax7-GFP^{high} cells can generate distinct daughter cell fates by asymmetric cell division. However, Pax7^{low} cells also express high levels of Myogenin, indicating that they are about to undergo differentiation. Therefore, Pax7^{high} and Pax7^{low} cells may reflect two stages of differentiation.

Molecular genetic studies in mice have established that a small subset of the satellite cell population are stem cells that are capable of reconstituting the satellite cell pool following transplantation, long-term self-renewal, and giving rise to committed myogenic progenitors through asymmetric apical-basal cell divisions [15]. Cre-LoxP lineage tracing in mice using Myf5-Cre and R26R-YFP alleles allows the discrimination between committed satellite myogenic cells that have expressed Myf5-Cre (YFP+), and a small subpopulation (<10%) of satellite stem cells that have never expressed Myf5-Cre (YFP–) [15]. Satellite stem cells also express high levels of Frizzled-7 (Fzd7), and signaling through the Wnt7a/Fzd7 planar cell polarity pathway drives the symmetric expansion of satellite stem cells to accelerate and augment muscle regeneration [16].

Under homeostatic conditions, the satellite cell pool is functionally heterogeneous when judged on the basis of proliferative history. Using tetracycline-inducible H2B-GFP mice, the Brack lab showed that the aged satellite cell compartment is composed of a label-retaining (LRCs) and non-label-retaining (non-LRCs) population [17]. Non-LRCs self-renew poorly, tend to differentiate and have a reduced regenerative capacity compared with LRCs which seem to rest in a more quiescent primitive state. In the intestinal system, it has been suggested that LRCs could display a dormant or reserve stem-cell population that is only active under conditions of injury [18].

Taken together, heterogeneity in the satellite cell population may indicate functional differenc-

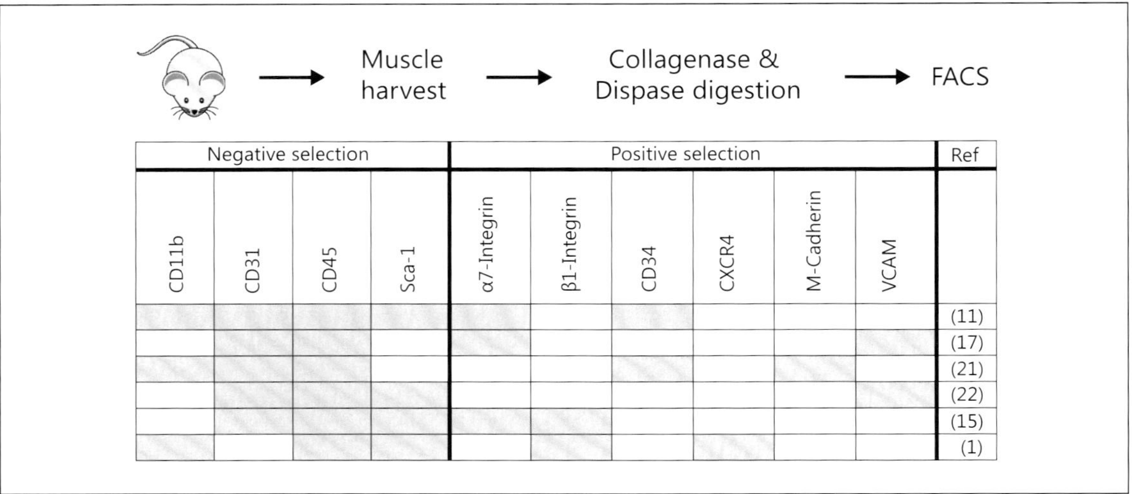

Fig. 3. Scheme of the experimental procedure for the isolation of murine satellite cells. The muscles are harvested and subjected to enzymatic and mechanical treatment. The myofiber-associated cells are then analyzed by fluorescence-activated cell sorting (FACS). Surface markers used for negative and positive selection for FACS analysis in different studies are highlighted in grey.

es or different stages in lineage specification of satellite cells (reviewed by Wagers and Conboy [19]). It was shown that satellite cells consist of both self-renewing stem cells and myogenic precursors [20]. The heterogeneity of satellite cells is evident in the differential expression of cell surface markers and differences in proliferation properties. There is so far no common approach to isolate satellite cells on the basis of cell surface marker expression (fig. 3); however, most purification methods employ a combination of antigens for negative selection (CD45, CD31, CD11b and Sca-1) as well as for positive selection (α_7-integrin, β_1-integrin, CXCR4, M-cadherin, VCAM) [1, 11, 15, 17, 21, 22]. Purity is generally assessed by Pax7 staining.

Origin of Satellite Cells

The origin of satellite cells has been analyzed in several studies [23–25]. Experiments based on quail-chick chimeras suggest that satellite cells of the trunk originate from the somites [23]. In more recent studies, this observation has been confirmed by applying molecular tracing techniques [25]. Using different transgenic reporter lines, the progenitors of trunk satellite cells have been identified in the central dermomyotome [25]. It was also shown that most limb muscle satellite cells are derived from hypaxial somites [25]. However, not all satellite cells have somatic origin; trunk and cranial satellite cells have separate genetic lineages [26] and some head muscles derive from prechordal mesoderm [27]. Other resources of satellite cells have been described, including bone marrow, vascular and hematopoietic lineages [24, 28], though the functional importance of these cells for physiological muscle growth and regeneration during postnatal life remains controversial.

The Molecular Basis of Satellite Cell Myogenesis

Satellite cells depend on the same regulatory components for determination and differentiation as the myogenic precursor cells in an embryo, including the paired box transcription factors Pax3 and Pax7 as well as the myogenic regulatory factors Myf5, MyoD and Myogenin (fig. 4) [29–31].

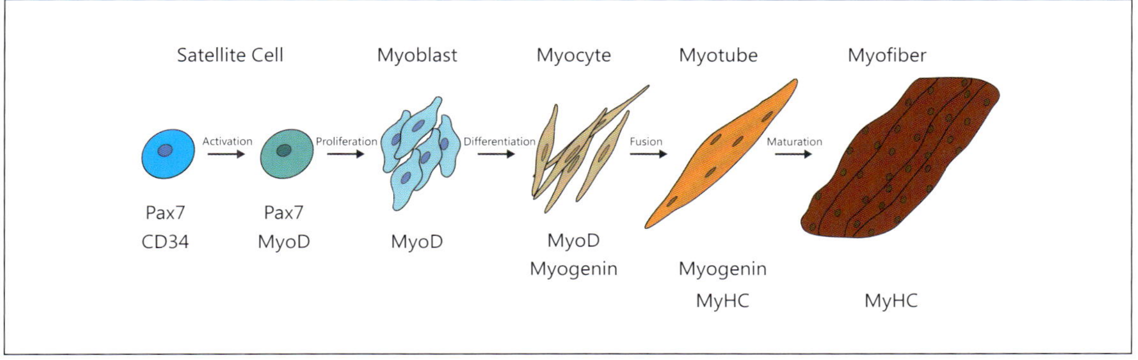

Fig. 4. Temporal expression of different components which are involved in the process of myogenesis.

Both Pax7 and to a lesser extent also Pax3 are expressed by quiescent satellite cells [14]. During activation of satellite cells, Pax7 and Pax3 are downregulated, as satellite cells start to differentiate [32]. Proportions of activated cells remain undifferentiated, retain Pax7 expression, and are thought to reconstitute the satellite cell pool. Pax7 seems to inhibit MyoD and Myogenin and thereby induces cells to reacquire a quiescent, undifferentiated state [32]. Both Pax3 and Pax7 simultaneously drive proliferation of satellite cell-derived myoblasts and keep them poised for differentiation in response to appropriate environmental cues [33]. In 2012, a genome-wide binding site analysis of Pax7 and Pax3 by Soleimani et al. [34] showed that both transcription factors bind identical DNA motifs and that they activate genes involved in muscle stem cell function. This analysis also reveals that Pax7 regulates a distinct set of genes implicated in proliferation and inhibition of differentiation and that this regulatory specificity can be explained by the different DNA-binding dynamics of the two transcription factors.

Further, it has been proposed that Pax7 mutant mice are completely devoid of satellite cells and that Pax7 is required for specifying the satellite cell lineage [35]. However, the function of Pax7 during satellite cell myogenesis was redefined [36]. It was demonstrated that satellite cells of Pax7 mutant mice show normal development, although their number was reduced. These data suggest that Pax7 is not required for the specification of myogenic satellite cells [32]. In order to study the function of Pax7 specifically in satellite cells, a conditional Pax7 allele was generated. This strain was used to conditionally ablate Pax7 function in myogenic cells when crossed with a Myf5-Cre mouse line. In these mice, muscle regeneration still occurs, although with reduced efficiency [36]. A gene-targeting approach to delete Pax7 during the proliferative response of adult satellite cells upon injury led to dramatically reduced satellite cell number and prevented muscle regeneration [21]. Furthermore, inactivation of Pax7 in Myf5-expressing cells after muscle damage leads to a complete arrest of muscle regeneration, indicating that replenishment of muscle stem cells from Myf5-expressing myogenic cells requires Pax7. Taken together, these studies suggest that Pax7 is not required for the specification of the muscle stem cell lineage, but rather for maintaining and renewing muscle stem cells. In addition, von Maltzahn et al. [37] have reported in 2013 that Pax7 expression is an absolute requirement for the normal function of satellite cells during regenerative myogenesis at any age.

Myf5 is expressed in quiescent cells [38]; MyoD is expressed as the cells become activated and subsequently differentiate under the control of Myogenin [39]. The myogenic differentiation program is completed with the expression of muscle-specific structural proteins like myosin heavy chain [7]. In addition, it was shown that specific microRNAs are highly expressed in quiescent satellite cells and are downregulated in the process of activation [22, 40].

Molecular Mechanism of Satellite Cell Aging

Aging is a complex process of functional decline in tissues and organs that leads to the deterioration of many body functions over the lifespan of an individual. Many of these age-related conditions exhibit a deregulation in tissue homeostasis and repair. In humans, aging is associated with a variety of health conditions, including cognitive decline, immune deficiency, cardiac dysfunction, atherosclerosis, wrinkling, increased bone fragility and muscle atrophy. Aging of muscle tissue includes a gradual decrease in skeletal muscle mass, strength and endurance which parallels with decreased regeneration potential leading to decreased physiological function [41]. This process is called sarcopenia.

Adult stem cells are considered to play a direct role in tissue regeneration, homeostasis and repair [42]. Impairment in the function of the tissue-specific stem cells may contribute to impairment in organ homeostasis during aging. Regarding satellite cells, there are discrepancies concerning their experimental quantification with age. Several studies showed a decline of muscle stem cells during aging [43–45]. In contrast, other groups did not see significant differences in satellite cell number between young and old muscle [46, 47]. However, it was shown that satellite cells from aged animals are activated with delay and have a reduced ability for proliferative expansion following muscle damage, indicating that a decline in satellite cell function may be the main reason for the impaired regeneration of skeletal muscle with age [44, 46, 48].

Several reports revealed that extrinsic cues have an impact on satellite cell function during aging, whereas the intrinsic capacity of satellite cells remains largely intact [45, 48, 49]. This hypothesis was supported by cell culture experiments using primary mouse satellite cells [48, 49]. When satellite cells of young mice were exposed to serum from old mice, a proportion of the cells lost their myogenic phenotype and acquired a fibroblastic appearance, suggesting that satellite cells adopt an altered fate by being exposed to systemic environmental factors from old mice. Moreover, the transcription factors MyoD and Pax7 were downregulated in satellite cells by an exposure to old serum [48, 49]. The conclusion that cell extrinsic factors can impair satellite cell function during aging was also supported by in vivo transplantation experiments. Aged muscle regenerates more efficiently when engrafted to a young host, but young muscle displays impaired regeneration when grafted into an aged host [50].

The Rando group [49] used an experimental system, termed parabiotic pairing, in which two mice are surgically connected to each other, thereby sharing a circulatory blood system. When a young mouse was attached to an old mouse (heterochronic parabiosis) the regenerative potential of the old satellite cells increased. This was not the case when two old mice were connected (isochronic parabiosis). The increased efficiency of regeneration in old satellite cells from heterochronic pairs also paralleled with an increased expression of the Notch signaling component Delta in the muscle progenitor cells [49]. Taken together, there are systemic factors which support regeneration efficiency of tissues in the young environment and/or which inhibit regeneration in old animals (fig. 5).

Besides circulation, also the local environment of the satellite cell has been related to the aging of the myogenic stem cell pool [17]. The local environment of a stem cell is represented by its niche, a specific anatomic location that regulates how the stem cell participates in tissue generation, maintenance and repair [51]. The niche is crucial for the maintenance of stem cell quiescence and the preservation of stem cell number and function, but also provides factors for activation [52]. The most important component of the satellite cell niche is the muscle fiber [53]. The Brack group [17] showed via label retention experiments that aged satellite cells spend less time in a quiescent state and have a reduced self-renewal potential. Furthermore, they demonstrate that in aged muscle fibers, fibroblast growth factor (FGF) 2 accumulates, acts as a mitogen and thereby promotes loss of satellite cell potency, eventually leading to a fall below a quorum of satellite cells that is required for an appropriate regenerative capacity.

The regenerative properties of satellite cells are controlled by various factors that regulate key molecular pathways including the Wnt, transforming growth factor-β (TGFβ), Notch, and NF-κB signaling pathways.

Wnt and TGFβ Signaling and Their Role in Satellite Cell Aging

Wnt signaling has been shown to be essential during embryogenesis and adult stem cell function [55]. Some publications divide the pathway into a canonical and a noncanonical signaling pathway. The canonical Wnt pathway involves the multifunctional protein β-catenin, whereas all Wnt signaling activities that are independent of β-catenin constitute different noncanonical Wnt signaling pathways [56].

Wnts are involved in embryonic myogenic induction in the paraxial mesoderm and also during muscle fiber differentiation [57, 58]. Work of the Rando group [48] showed that Wnt signaling also has a critical role in stem cell aging. Analysis of components of the Wnt signaling cascade indicated higher canonical Wnt signaling activity in aged satellite cells compared to young. When adding additional Wnt3a to satellite cell cultures exposed to young serum, there was an increased myogenic-to-fibrogenic conversion of satellite cells. Vice versa, when satellite cells were exposed

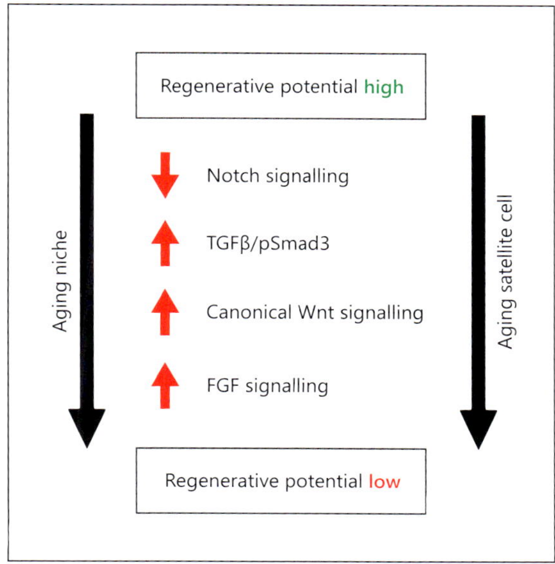

Fig. 5. Model illustrating the age-dependent regeneration potential of satellite cells in response to intrinsic factors (cell age) and extrinsic factors (niche age). During aging, satellite cells acquire intrinsic changes and are exposed to a modified local and systemic environment. The molecular pathways (Notch, TGFβ/Smad3, Wnt and fibroblast growth factor, FGF) that mediate the extrinsic age-dependent effects are summarized in the center. Some of the signaling pathways show an increased activity during aging (e.g. TGFβ), whereas other signaling pathways show a decreased activity (e.g. Delta). Modified from Silva and Conboy [54].

to old serum, the degree of lineage conversion was reduced by adding a Wnt inhibitor – the Frizzled-related protein 3 [48]. This implies that the effect of the environment is partly mediated by increased canonical Wnt signaling in aged mice, acting on satellite cells to lose their myogenic lineage and become fibroblast-like cells.

In contrast, noncanonical Wnt7a signaling has been shown to drive symmetric expansion of satellite cells through the planar cell polarity pathway, leading to accelerated and augmented muscle regeneration [16]. Recently, satellite cell-specific expression of the extracellular matrix glycoprotein fibronectin has been found to remodel the stem cell niche in muscle to regulate Wnt7a signaling and satellite stem cell expansion, again stressing the importance of the niche for satellite cell function [59]. Fibronectin binds to Syndecan-4, which forms a coreceptor complex with Fzd7 in satellite cells, stimulating the ability of Wnt7a to induce the symmetric expansion of satellite cells. Furthermore, Wnt7a regulates muscle growth, acting through the Akt/mTOR pathway in differentiated muscle fibers to induce hypertrophy [60]. Thereby, Wnt7a seems to couple tissue size and mass to the stem cell pool and vice versa. Ultimately, Wnt7a treatment efficiently induces satellite cell expansion and myofiber hypertrophy in treated muscles in mdx mice (see below), leading to a significant increase in muscle strength and a reduction in the level of contractile damage [61]. Additionally, this study also indicated that the function of the Wnt7a/Fzd7 signaling pathway in skeletal muscle is conserved between humans and mice, making Wnt7a a promising candidate for the development of molecular therapies for treatment of Duchene muscular dystrophy (DMD).

The TGF family of cytokines exerts a wide range of biological effects on a large variety of cell types, for example it regulates cell growth, differentiation, matrix production and apoptosis [62]. Binding of TGFs to their receptors activates phosphorylation of Smad proteins, leading to their nuclear translocation and activation of target genes [62]. Among TGF members, TGFβ acts as inhibitor for satellite cells in their progression through the myogenic lineage [63]. Excessive TGFβ/phospho-Smad3 in aged satellite cells fosters the decline in Notch activation and leads to accumulation of cyclin-dependent kinase inhibitors in satellite cells, thus preventing regeneration of damaged skeletal muscle [64].

Notch Signaling: Control of Regenerative Competence of Satellite Cells

The role of Notch signaling has been described in several biological processes, including tissue formation during embryonic development [65]. Notch controls developmental patterning, cell fate decisions, as well as proliferation and maintenance of progenitor cells [66]. The receptor Notch and its ligand Delta are transmembrane proteins. Upon binding of the ligand to the receptor, the intracellular portion of Notch is cleaved, translocates to the nucleus where it regulates the expression of several target genes as a transcription factor.

In the last years, it became apparent that Notch signaling is activated in postnatal myogenic lineage progression and that it has a role in satellite cell aging [46, 67]. In damaged muscles, the Notch pathway is activated by upregulation of Delta and regulates the transition from the quiescent cell to proliferating precursor cells. Ectopic expression of Notch promotes proliferation of satellite cells and attenuates myogenic differentiation of muscle progenitor cells. Conversely, blocking the Notch pathway leads to an upregulation of muscle differentiation genes and reduced proliferation of satellite cells. Upon injury, expression of the ligand Delta is reduced in aged compared to young satellite cells. The impaired activation of Delta is associated with a decreased proliferation and expansion of muscle progenitor cells and an inefficient regeneration [46]. In addition, muscle regeneration was inhibited when the Notch pathway was blocked at the site of injury. In contrast, ectopic induction of Notch signaling improved regeneration of old muscles.

NF-κB Signaling and Its Role in Satellite Cell Aging

The NF-κB pathway is mediating cellular signaling in response to several stimuli including stress and cytokines [68]. Wager's group [unpubl. data] analyzed gene expression of old satellite cells versus old satellite cells that were rejuvenated by heterochronic parabiosis. This screen identified differentially expressed components of the NF-κB pathway.

NF-κB activity negatively regulates muscle stem cell function leading to skeletal muscle atrophy. Analyses on mutant mice support that the

NF-κB pathway functions as an inhibitor of skeletal myogenesis and muscle regeneration [69, 70]. Furthermore, p65, a subunit of NF-κB, is involved in suppressing MyoD mRNA at the post-transcriptional level and ultimately reduced skeletal muscle differentiation [70]. As complete knockout of p65 results in embryonic lethality [71], stem cells derived from skeletal muscles have been analyzed in $p65^{+/-}$ mice, which show enhanced cell proliferation and myogenic differentiation [72]. In addition, it has been shown that lowering basal levels of NF-κB activity by gene knockout or pharmacologic inhibition of NF-κB increased the capacity of muscle stem cells to engraft and muscle regeneration after implantation into dystrophic mouse skeletal muscles [72].

Telomere Shortening in Satellite Cells

DMD is a devastating X-linked muscle degenerative disorder in which loss of dystrophin causes muscle weakness and fragility [73]. DMD patients suffer from progressive loss of muscle function, leading to paralysis and premature death [74]. The mouse model (mdx mouse) with the same muscle structural defect as DMD patients (a mutation in the dystrophin gene) exhibits only mild muscle weakness and little reduction in lifespan [73, 74]. Moreover, the skeletal muscles of the mdx mouse are capable of repeated cycles of degeneration/regeneration and their regenerative capacity does not appear to be exhausted in contrast to DMD patients who exhibit diminishing regeneration capacity.

In studies by Blau et al. [75] in 1983, it was shown that myoblasts from DMD patients have an impaired replicative potential. This finding and a relevant correlation with telomere shortening have been reported to be a common feature of dystrophic human muscle cells with increasing age by the group of Butler-Browne [76]. Humans have relatively short telomeres (~8 kb), whereas the telomeres from laboratory mice are much longer (40 kb). The Blau lab postulated that the mild dystrophic phenotype in mdx mice results from the ability of their satellite cells to meet the high regenerative demand of dystrophic tissues throughout their lifetime, due to their longer telomeres. To test this hypothesis, they generated mice lacking both dystrophin and telomerase activity (mdx/mTR mice) and demonstrated that these mice recapitulate the severe phenotype of DMD patients [77]. This includes markedly elevated creatine kinase levels, severe loss of muscle force, kyphosis, muscle membrane disruption, and progressive wasting of limb and diaphragm muscles, culminating in early death [77].

Under normal conditions, skeletal muscle is relatively quiescent when compared to high-turnover tissues such as the blood and most epithelia. In DMD patients, however, there are elevated levels of muscle damage and therefore proliferation rates of satellite cells are increased to accomplish regeneration. When testing the functional reserve of satellite cells of dystrophin mutant mice with short telomeres compared to mice with long telomeres, satellite cells of dystrophic animals exhibit reduced proliferative capacity in vitro and in vivo [77]. Importantly, transplantation of satellite cells with long telomeres into mdx/mTR mice ameliorated the dystrophic phenotype, suggesting that DMD progression results from a cell autonomous failure of satellite cells to regenerate dystrophic muscles. These findings support the hypothesis that in humans DMD is initiated by the dystrophin structural genetic defect but develops progressively due to an exhaustion of the stem cell pool caused by telomere shortening.

Apoptosis of Satellite Cells in the Process of Aging

Apoptosis, or programmed cell death, was described in several biological processes to ensure homeostasis of tissues by the elimination of compromised or superfluous cells, especially during development. However, there is also emerging evidence that apoptosis can contribute to the depletion of stem cells and tissue dystrophy during aging [78]. There are a number of cellular mechanisms through which apoptosis can be induced including DNA damage, growth factors and cytokines [79]. Apoptotic stimuli comprise both extrinsic and intrinsic signals that are produced following cellular stress. A family of proteins known as caspases is typically activated during apoptosis [80]. These proteins cleave key cellular components that are required for normal cellular function including structural proteins in the cytoskeleton and nuclear proteins such as DNA repair enzymes.

Recent studies showed that alterations in apoptosis contribute to organismal aging [81]. For example, studies in adult animals revealed that apoptosis is involved in the loss of muscle nuclei during acute disuse atrophy, and that caspase-3-dependent pathways play an important role in this process [82]. Moreover, increased apoptosis was observed in splenocytes and thymocytes obtained from aged rats compared to young ones [83]. In addition, studies on deletion of PUMA (a p53 target gene inducing apoptosis) revealed the first experimental evidence that apoptosis contributes to tissue atrophy and lifespan shortening in the context of telomere dysfunction [78]. Taken together, increased rate of apoptosis could contribute to tissue degeneration and aging.

Besides diminished satellite cell proliferation, aberrant differentiation and failure of self-renewal, apoptosis has been proposed as an additional mechanism contributing to the age-dependent depletion of satellite cells [84]. Recent studies indicate that a greater proportion of satellite cells of old rats underwent programmed cell death, when compared to young satellite cells [85]. Furthermore, satellite cells from old muscle contained less BCL2, a molecular inhibitor of apoptotic cell death [85].

Epigenetic Regulation of Myogenic Stem Cells during Aging

Several studies suggest that stem cell aging involves changes in the epigenome [86–88]. A detailed analysis of histone methylation patterns of young and old quiescent muscle stem cells revealed that there is an accumulation of repressed chromatin domains in aged quiescent satellite cells, which could account for their decreased functionality [86]. Changes in the epigenetic characteristics of satellite cells lead to different events including fate conversion and cell division during myogenesis and aging.

During muscle aging, the proportion of fibrous connective tissue increases [89]. This process is driven by altered muscle stem cell differentiation. The fate conversion from myogenic to fibroblast-like cells was shown to be regulated by altered canonical Wnt signaling during aging [48]. Mutational analysis in mice showed that one of the Wnt components, BCL9, has a functional role in myogenic lineage progression in response to Wnt signaling [90]. In addition, it was shown that the BCL9 complex is binding to methylated lysine of histone 3 [91]. Therefore, the complex is not only a transcriptional coactivator, but is also a binding component for activated genes that have this epigenetic mark. This provides a link of Wnt regulator components, lineage progression and epigenetic regulation [90].

During the asymmetric division of satellite cells, daughter cells have two different fates; one returns to a quiescent state to refill the stem cell pool (Pax7$^+$ cells) and the other one will enter the differentiation program (MyoD$^+$ and Myf5$^+$ cells). This process is regulated by epigenetic changes that permit a coordinated expression of a subset of genes in the distinct daughter cells. The chromatin changes coordinate lineage determination by repression and de-repression of lineage-specific genes [92, 93].

Several transcription factors have been shown to be involved in chromatin remodeling during myogenesis including MyoD and Pax7 [94, 95]. It has been demonstrated that MyoD remodels chromatin of silent loci and thereby activates muscle-specific genes [94]. Pax7 was shown to enforce satellite cell commitment by chromatin remodeling. Pax7 is associated with a histone methyltransferase complex that modifies histone 3. This complex interacts with the Myf5 locus resulting in methylation of the surrounding chromatin and thereby allowing the activation of gene expression [95]. Furthermore, in 2012, Kawabe et al. [96] demonstrated that the methyltransferase Carm1 specifically methylates Pax7, and this induces the satellite cell to enter the myogenic differentiation by asymmetric cell division.

Molecular Mechanism of Cardiac Muscle Aging and Disease

Aging-induced molecular changes in the cardiac muscle lead to heart disease and reduced heart function. Cardiac muscle is formed by cardiomyocytes, which are binucleated in the adult murine heart and in contrast to skeletal muscle characterized by a high amount of mitochondria [97]. Cardiac muscle has a limited capacity to respond to stress and injury; it displays a low capacity of regeneration due to the poor ability to enter the cell cycle and undergo cell division (reviewed by Kikuchi and Poss [98]).

Injury-induced stress and aging of cardiac muscle leads to different molecular and cellular events including cardiomyocyte hypertrophy, apoptosis, oxidative stress, telomere erosion and cardiomyocyte remodeling [99–102].

Cardiac Hypertrophy during Aging

Heart failure in old patients is often associated with cardiac hypertrophy [99]. Using a parabiosis model (see above), Loffredo et al. [100] showed that the aging-induced hypertrophy can be reversed by exposure to a young environment. They demonstrate that aging-related hypertrophy is modulated at least in part by growth differentiation factor 11 (GDF11). GDF11 is a member of the activin/TGFβ family of growth factors and activates the Smad2/3 pathway. The authors demonstrate that the expression of this factor is reduced in the circulation of aged compared to young mice. In vitro and in vivo studies show that administration of GDF11 has a direct anti-hypertrophic effect and leads to changes in molecular markers associated with cardiac hypertrophy.

De-Differentiation as a Mechanism of Cardiac Muscle Aging

The group of Braun [103] observed an age-dependent increase in de-differentiation in cardiomyocytes in patients with dilative cardiomyopathy. De-differentiation of cardiomyocytes is a process of cell remodeling that occurs during acute and chronic diseases. In order to identify factors that trigger the process of de-differentiation, the Braun group set up a proteomic screen. They identified factors which were differentially expressed in old compared to young serum. One of these factors is a member of the Oncostatin M (OSM) signaling pathway. OSM is a member of the cytokine IL6 family, which interacts with both OSM-R/gp130 and LIF-R/gp130 receptor, and it was shown to display both pro-inflammatory and anti-inflammatory activities [104, 105]. OSM shows a strong upregulation in patients with dilatative cardiomyopathy. In vitro analysis showed that OSM induces de-differentiation of cultured myocytes to an immature state. This observation was confirmed by systemic administration of OSM into mice. This in vivo experiment showed that OSM is sufficient to induce partial de-differentiation of cardiomyocytes. Further, the group demonstrated that inhibition of OSM-mediated cardiomyocyte de-differentiation impairs survival after acute myocardial infarction and improves survival in chronic dilative cardiomyopathy.

Telomere Dysfunction and Cardiac Failure

Telomere dysfunction has been proposed to play a critical role in cancer progression and aging. Late-generation telomerase-deficient mice show severe phenotype including abnormalities of the hematopoietic system and atrophy of the small intestine [106]. It has been further demonstrated that these mice show pathological cardiac remodeling as a consequence of impaired cardiomyocyte regeneration and increased cardiomyocyte apoptosis [107]. The dystrophin mouse model (mdx) has been also used to evaluate the effect of telomere shortening on cardiac function [101]. Mice with the mdx mutation alone have a relatively mild phenotype [77]. However, mice lacking dystrophin with shortened telomeres (mdx/mTR mice) recapitulate the severe phenotype of DMD patients [77, 101]. Further studies show that these double mutants also develop functional cardiac defects like in DMD patients [101]. On a molecular level, these cardiac defects are associated with mitochondrial fragmentation, telomere erosion and increased oxidative stress leading to accelerated organ failure. In order to understand whether oxidative stress causes the pathogenesis of the hearts in the compound mutant, the authors subjected the mice to antioxidants. Both cardiac morphology and function were rescued by the administration of the antioxidant and prolonged the lifespan of the double mutant. This suggests that oxidative stress is a critical factor in the pathogenesis of cardiac failure and supports earlier studies which proposed that there is a correlation between telomere shortening and oxidative stress [108, 109].

Conclusions and Perspectives

Our knowledge about aging of satellite cells is growing rapidly. This was underlined by recent talks at the 5th Else Kröner-Fresenius Symposium. It is becoming clear that the dysfunction of aged muscle progenitors is to some extent induced by systemic regulators, and these alterations appear to be partially reversible in satellite cells when reexposed to a young environment. To reverse age-associated impairments in satellite cell function, specific causes of muscle progenitor cell aging that are present in the cellular environment as well as cell intrinsically have to be identified. Age-dependent processes in satellite cells also involve genetic and epigenetic modifications resulting in changes in gene expression. Aging-associated epigenetic alterations may also affect the stem cell niche or the composition of systemic-acting factors in the blood serum. The analysis of gene expression profiles of (young and old) satellite cells led to the identification of new signaling pathways which mediate the age-dependent decrease in satellite cell functionality. In the future, it will be important to elucidate reversibility and prevention of molecular mechanisms of satellite cell aging. This knowledge may offer novel therapeutic approaches for prevention and treatment of age-related muscle atrophy.

References

1 Cerletti M, Jurga S, Witczak CA, Hirshman MF, Shadrach JL, Goodyear LJ, Wagers AJ: Highly efficient, functional engraftment of skeletal muscle stem cells in dystrophic muscles. Cell 2008;134:37–47.

2 Mauro A: Satellite cell of skeletal muscle fibers. J Biophys Biochem Cytol 1961;9:493–495.

3 Schultz E: Fine structure of satellite cells in growing skeletal muscle. Am J Anat 1976;147:49–70.

4 Muir AR, Kanji AH, Allbrook D: The structure of the satellite cells in skeletal muscle. J Anat 1965;99:435–444.

5 Schultz E, McCormick KM: Skeletal muscle satellite cells. Rev Physiol Biochem Pharmacol 1994;123:213–257.

6 Chargé SB, Rudnicki MA: Cellular and molecular regulation of muscle regeneration. Physiol Rev 2004;84:209–238.

7 Collins CA, Olsen I, Zammit PS, Heslop L, Petrie A, Partridge TA, Morgan JE: Stem cell function, self-renewal, and behavioral heterogeneity of cells from the adult muscle satellite cell niche. Cell 2005;122:289–301.

8 Collins CA, Partridge TA: Self-renewal of the adult skeletal muscle satellite cell. Cell Cycle 2005;4:1338–1341.

9 Hawke TJ, Garry DJ: Myogenic satellite cells: physiology to molecular biology. J Appl Physiol 2001;91:534–551.

10 Moss FP, Leblond CP: Satellite cells as the source of nuclei in muscles of growing rats. Anat Rec 1971;170:421–435.

11 Sacco A, Doyonnas R, Kraft P, Vitorovic S, Blau HM: Self-renewal and expansion of single transplanted muscle stem cells. Nature 2008;456:502–506.

12 Rocheteau P, Gayraud-Morel B, Siegl-Cachedenier I, Blasco MA, Tajbakhsh S: A subpopulation of adult skeletal muscle stem cells retains all template DNA strands after cell division. Cell 2012; 148:112–125.
13 Montarras D, Morgan J, Collins C, Relaix F, Zaffran S, Cumano A, Partridge T, Buckingham M: Direct isolation of satellite cells for skeletal muscle regeneration. Science 2005;309:2064–2067.
14 Relaix F, Montarras D, Zaffran S, Gayraud-Morel B, Rocancourt D, Tajbakhsh S, Mansouri A, Cumano A, Buckingham M: Pax3 and Pax7 have distinct and overlapping functions in adult muscle progenitor cells. J Cell Biol 2006;172:91–102.
15 Kuang S, Kuroda K, Le Grand F, Rudnicki MA: Asymmetric self-renewal and commitment of satellite stem cells in muscle. Cell 2007;129:999–1010.
16 Le Grand F, Jones AE, Seale V, Scime A, Rudnicki MA: Wnt7a activates the planar cell polarity pathway to drive the symmetric expansion of satellite stem cells. Cell Stem Cell 2009;4:535–547.
17 Chakkalakal JV, Jones KM, Basson MA, Brack AS: The aged niche disrupts muscle stem cell quiescence. Nature 2012;490:355–360.
18 Buczacki SJ, Zecchini HI, Nicholson AM, Russell R, Vermeulen L, Kemp R, Winton DJ: Intestinal label-retaining cells are secretory precursors expressing Lgr5. Nature 2013;495:65–69.
19 Wagers AJ, Conboy IM: Cellular and molecular signatures of muscle regeneration: current concepts and controversies in adult myogenesis. Cell 2005; 122:659–667.
20 Sherwood RI, Christensen JL, Conboy IM, Conboy MJ, Rando TA, Weissman IL, Wagers AJ: Isolation of adult mouse myogenic progenitors: functional heterogeneity of cells within and engrafting skeletal muscle. Cell 2004;119:543–554.
21 Gunther S, Kim J, Kostin S, Lepper C, Fan CM, Braun T: Myf5-positive satellite cells contribute to Pax7-dependent long-term maintenance of adult muscle stem cells. Cell Stem Cell 2013;13:590–601.
22 Cheung TH, Quach NL, Charville GW, Liu L, Park L, Edalati A, Yoo B, Hoang P, Rando TA: Maintenance of muscle stem-cell quiescence by microRNA-489. Nature 2012;482:524–528.
23 Armand O, Boutineau AM, Mauger A, Pautou MP, Kieny M: Origin of satellite cells in avian skeletal muscles. Arch Anat Microsc Morphol Exp 1983;72: 163–181.
24 Asakura A, Rudnicki MA: Side population cells from diverse adult tissues are capable of in vitro hematopoietic differentiation. Exp Hematol 2002;30: 1339–1345.
25 Schienda J, Engleka KA, Jun S, Hansen MS, Epstein JA, Tabin CJ, Kunkel LM, Kardon G: Somitic origin of limb muscle satellite and side population cells. Proc Natl Acad Sci USA 2006;103:945–950.
26 Harel I, Nathan E, Tirosh-Finkel L, Zigdon H, Guimaraes-Camboa N, Evans SM, Tzahor E: Distinct origins and genetic programs of head muscle satellite cells. Dev Cell 2009;16:822–832.
27 Lu JR, Bassel-Duby R, Hawkins A, Chang P, Valdez R, Wu H, Gan L, Shelton JM, Richardson JA, Olson EN: Control of facial muscle development by MyoR and capsulin. Science 2002;298: 2378–2381.
28 Gussoni E, Soneoka Y, Strickland CD, Buzney EA, Khan MK, Flint AF, Kunkel LM, Mulligan RC: Dystrophin expression in the mdx mouse restored by stem cell transplantation. Nature 1999;401: 390–394.
29 Maroto M, Reshef R, Munsterberg AE, Koester S, Goulding M, Lassar AB: Ectopic Pax-3 activates MyoD and Myf-5 expression in embryonic mesoderm and neural tissue. Cell 1997;89:139–148.
30 Rudnicki MA, Schnegelsberg PN, Stead RH, Braun T, Arnold HH, Jaenisch R: MyoD or Myf-5 is required for the formation of skeletal muscle. Cell 1993;75: 1351–1359.
31 Zammit PS, Partridge TA, Yablonka-Reuveni Z: The skeletal muscle satellite cell: the stem cell that came in from the cold. J Histochem Cytochem 2006;54: 1177–1191.
32 Olguin HC, Olwin BB: Pax-7 up-regulation inhibits myogenesis and cell cycle progression in satellite cells: a potential mechanism for self-renewal. Dev Biol 2004;275:375–388.
33 Collins CA, Gnocchi VF, White RB, Boldrin L, Perez-Ruiz A, Relaix F, Morgan JE, Zammit PS: Integrated functions of Pax3 and Pax7 in the regulation of proliferation, cell size and myogenic differentiation. PLoS One 2009;4:e4475.
34 Soleimani VD, Punch VG, Kawabe Y, Jones AE, Palidwor GA, Porter CJ, Cross JW, Carvajal JJ, Kockx CE, van IJcken WF, Perkins TJ, Rigby PW, Grosveld F, Rudnicki MA: Transcriptional dominance of Pax7 in adult myogenesis is due to high-affinity recognition of homeodomain motifs. Dev Cell 2012;22:1208–1220.
35 Seale P, Sabourin LA, Girgis-Gabardo A, Mansouri A, Gruss P, Rudnicki MA: Pax7 is required for the specification of myogenic satellite cells. Cell 2000;102: 777–786.
36 Oustanina S, Hause G, Braun T: Pax7 directs postnatal renewal and propagation of myogenic satellite cells but not their specification. EMBO J 2004;23: 3430–3439.
37 von Maltzahn J, Jones AE, Parks RJ, Rudnicki MA: Pax7 is critical for the normal function of satellite cells in adult skeletal muscle. Proc Natl Acad Sci USA 2013;110:16474–16479.
38 Beauchamp JR, Heslop L, Yu DS, Tajbakhsh S, Kelly RG, Wernig A, Buckingham ME, Partridge TA, Zammit PS: Expression of CD34 and Myf5 defines the majority of quiescent adult skeletal muscle satellite cells. J Cell Biol 2000;151:1221–1234.
39 Yablonka-Reuveni Z, Rivera AJ: Temporal expression of regulatory and structural muscle proteins during myogenesis of satellite cells on isolated adult rat fibers. Dev Biol 1994;164:588–603.
40 Boutet SC, Cheung TH, Quach NL, Liu L, Prescott SL, Edalati A, Iori K, Rando TA: Alternative polyadenylation mediates microRNA regulation of muscle stem cell function. Cell Stem Cell 2012; 10:327–336.
41 Degens H: Age-related skeletal muscle dysfunction: causes and mechanisms. J Musculoskelet Neuronal Interact 2007; 7:246–252.
42 Rando TA: Stem cells, ageing and the quest for immortality. Nature 2006; 441:1080–1086.
43 Gibson MC, Schultz E: Age-related differences in absolute numbers of skeletal muscle satellite cells. Muscle Nerve 1983;6:574–580.
44 Schultz E, Lipton BH: Skeletal muscle satellite cells: changes in proliferation potential as a function of age. Mech Ageing Dev 1982;20:377–383.
45 Shefer G, Van de Mark DP, Richardson JB, Yablonka-Reuveni Z: Satellite-cell pool size does matter: defining the myogenic potency of aging skeletal muscle. Dev Biol 2006;294:50–66.
46 Conboy IM, Conboy MJ, Smythe GM, Rando TA: Notch-mediated restoration of regenerative potential to aged muscle. Science 2003;302:1575–1577.

47 Nnodim JO: Satellite cell numbers in senile rat levator ani muscle. Mech Ageing Dev 2000;112:99–111.
48 Brack AS, Conboy MJ, Roy S, Lee M, Kuo CJ, Keller C, Rando TA: Increased Wnt signaling during aging alters muscle stem cell fate and increases fibrosis. Science 2007;317:807–810.
49 Conboy IM, Conboy MJ, Wagers AJ, Girma ER, Weissman IL, Rando TA: Rejuvenation of aged progenitor cells by exposure to a young systemic environment. Nature 2005;433:760–764.
50 Carlson BM, Faulkner JA: Muscle transplantation between young and old rats: age of host determines recovery. Am J Physiol 1989;256:C1262–C1266.
51 Scadden DT: The stem-cell niche as an entity of action. Nature 2006;441:1075–1079.
52 Brack AS, Rando TA: Tissue-specific stem cells: lessons from the skeletal muscle satellite cell. Cell Stem Cell 2012;10:504–514.
53 Gopinath SD, Rando TA: Stem cell review series: aging of the skeletal muscle stem cell niche. Aging Cell 2008;7:590–598.
54 Silva H, Conboy IM: Aging and stem cell renewal. StemBook (Internet). Cambridge, Harvard Stem Cell Institute, 2008.
55 Klaus A, Birchmeier W: Wnt signalling and its impact on development and cancer. Nat Rev Cancer 2008;8:387–398.
56 Rao TP, Kuhl M: An updated overview on Wnt signaling pathways: a prelude for more. Circ Res 2010;106:1798–1806.
57 Anakwe K, Robson L, Hadley J, Buxton P, Church V, Allen S, Hartmann C, Harfe B, Nohno T, Brown AM, Evans DJ, Francis-West P: Wnt signalling regulates myogenic differentiation in the developing avian wing. Development 2003;130:3503–3514.
58 Tajbakhsh S, Borello U, Vivarelli E, Kelly R, Papkoff J, Duprez D, Buckingham M, Cossu G: Differential activation of Myf5 and MyoD by different Wnts in explants of mouse paraxial mesoderm and the later activation of myogenesis in the absence of Myf5. Development 1998;125:4155–4162.
59 Bentzinger CF, Wang YX, von Maltzahn J, Soleimani VD, Yin H, Rudnicki MA: Fibronectin regulates Wnt7a signaling and satellite cell expansion. Cell Stem Cell 2013;12:75–87.
60 von Maltzahn J, Bentzinger CF, Rudnicki MA: Wnt7a-Fzd7 signalling directly activates the Akt/mTOR anabolic growth pathway in skeletal muscle. Nat Cell Biol 2012;14:186–191.
61 von Maltzahn J, Renaud JM, Parise G, Rudnicki MA: Wnt7a treatment ameliorates muscular dystrophy. Proc Natl Acad Sci USA 2012;109:20614–20619.
62 Heldin CH, Miyazono K, ten Dijke P: TGF-beta signalling from cell membrane to nucleus through SMAD proteins. Nature 1997;390:465–471.
63 Allen RE, Boxhorn LK: Inhibition of skeletal muscle satellite cell differentiation by transforming growth factor-beta. J Cell Physiol 1987;133:567–572.
64 Carlson ME, Conboy MJ, Hsu M, Barchas L, Jeong J, Agrawal A, Mikels AJ, Agrawal S, Schaffer DV, Conboy IM: Relative roles of TGF-beta1 and Wnt in the systemic regulation and aging of satellite cell responses. Aging Cell 2009;8:676–689.
65 Baron M: An overview of the Notch signalling pathway. Semin Cell Dev Biol 2003;14:113–119.
66 Yoon K, Gaiano N: Notch signaling in the mammalian central nervous system: insights from mouse mutants. Nat Neurosci 2005;8:709–715.
67 Conboy IM, Rando TA: The regulation of Notch signaling controls satellite cell activation and cell fate determination in postnatal myogenesis. Dev Cell 2002; 3:397–409.
68 Gilmore TD: Introduction to NF-kappaB: players, pathways, perspectives. Oncogene 2006;25:6680–6684.
69 Cai D, Frantz JD, Tawa NE Jr, Melendez PA, Oh BC, Lidov HG, Hasselgren PO, Frontera WR, Lee J, Glass DJ, Shoelson SE: IKKbeta/NF-kappaB activation causes severe muscle wasting in mice. Cell 2004;119:285–298.
70 Guttridge DC, Mayo MW, Madrid LV, Wang CY, Baldwin AS Jr: NF-kappaB-induced loss of MyoD messenger RNA: possible role in muscle decay and cachexia. Science 2000;289:2363–2366.
71 Beg AA, Sha WC, Bronson RT, Ghosh S, Baltimore D: Embryonic lethality and liver degeneration in mice lacking the RelA component of NF-kappa B. Nature 1995;376:167–170.
72 Lu A, Proto JD, Guo L, Tang Y, Lavasani M, Tilstra JS, Niedernhofer LJ, Wang B, Guttridge DC, Robbins PD, Huard J: NF-kappaB negatively impacts the myogenic potential of muscle-derived stem cells. Mol Ther 2012;20:661–668.
73 Hoffman EP, Brown RH Jr, Kunkel LM: Dystrophin: the protein product of the Duchenne muscular dystrophy locus. Cell 1987;51:919–928.
74 Emery AE: The muscular dystrophies. Lancet 2002;359:687–695.
75 Blau HM, Webster C, Chiu CP, Guttman S, Chandler F: Differentiation properties of pure populations of human dystrophic muscle cells. Exp Cell Res 1983;144:495–503.
76 Decary S, Hamida CB, Mouly V, Barbet JP, Hentati F, Butler-Browne GS: Shorter telomeres in dystrophic muscle consistent with extensive regeneration in young children. Neuromuscul Disord 2000;10:113–120.
77 Sacco A, Mourkioti F, Tran R, Choi J, Llewellyn M, Kraft P, Shkreli M, Delp S, Pomerantz JH, Artandi SE, Blau HM: Short telomeres and stem cell exhaustion model Duchenne muscular dystrophy in mdx/mTR mice. Cell 2010;143: 1059–1071.
78 Sperka T, Song Z, Morita Y, Nalapareddy K, Guachalla LM, Lechel A, Begus-Nahrmann Y, Burkhalter MD, Mach M, Schlaudraff F, Liss B, Ju Z, Speicher MR, Rudolph KL: Puma and p21 represent cooperating checkpoints limiting self-renewal and chromosomal instability of somatic stem cells in response to telomere dysfunction. Nat Cell Biol 2012;14:73–79.
79 Li H, Yuan J: Deciphering the pathways of life and death. Curr Opin Cell Biol 1999;11:261–266.
80 Budihardjo I, Oliver H, Lutter M, Luo X, Wang X: Biochemical pathways of caspase activation during apoptosis. Annu Rev Cell Dev Biol 1999;15:269–290.
81 Zhang Y, Herman B: Ageing and apoptosis. Mech Ageing Dev 2002;123:245–260.
82 Dupont-Versteegden EE: Apoptosis in muscle atrophy: relevance to sarcopenia. Exp Gerontol 2005;40:473–481.
83 Kapasi AA, Singhal PC: Aging splenocyte and thymocyte apoptosis is associated with enhanced expression of p53, bax, and caspase-3. Mol Cell Biol Res Commun 1999;1:78–81.
84 Jones DL, Rando TA: Emerging models and paradigms for stem cell ageing. Nat Cell Biol 2011;13:506–512.
85 Jejurikar SS, Henkelman EA, Cederna PS, Marcelo CL, Urbanchek MG, Kuzon WM Jr: Aging increases the susceptibility of skeletal muscle derived satellite cells to apoptosis. Exp Gerontol 2006; 41:828–836.

86 Liu L, Cheung TH, Charville GW, Hurgo BM, Leavitt T, Shih J, Brunet A, Rando TA: Chromatin modifications as determinants of muscle stem cell quiescence and chronological aging. Cell Rep 2013;4:189–204.
87 Liu L, Rando TA: Manifestations and mechanisms of stem cell aging. J Cell Biol 2011;193:257–266.
88 Oberdoerffer P, Michan S, McVay M, Mostoslavsky R, Vann J, Park SK, Hartlerode A, Stegmuller J, Hafner A, Loerch P, Wright SM, Mills KD, Bonni A, Yankner BA, Scully R, Prolla TA, Alt FW, Sinclair DA: SIRT1 redistribution on chromatin promotes genomic stability but alters gene expression during aging. Cell 2008;135:907–918.
89 Goldspink G, Fernandes K, Williams PE, Wells DJ: Age-related changes in collagen gene expression in the muscles of mdx dystrophic and normal mice. Neuromuscul Disord 1994;4:183–191.
90 Brack AS, Murphy-Seiler F, Hanifi J, Deka J, Eyckerman S, Keller C, Aguet M, Rando TA: BCL9 is an essential component of canonical Wnt signaling that mediates the differentiation of myogenic progenitors during muscle regeneration. Dev Biol 2009;335:93–105.
91 Fiedler M, Sanchez-Barrena MJ, Nekrasov M, Mieszczanek J, Rybin V, Muller J, Evans P, Bienz M: Decoding of methylated histone H3 tail by the Pygo-BCL9 Wnt signaling complex. Mol Cell 2008;30:507–518.
92 Fischle W, Wang Y, Allis CD: Histone and chromatin cross-talk. Curr Opin Cell Biol 2003;15:172–183.
93 Vaquero A, Loyola A, Reinberg D: The constantly changing face of chromatin. Sci Aging Knowledge Environ 2003; 2003:RE4.
94 Gerber AN, Klesert TR, Bergstrom DA, Tapscott SJ: Two domains of MyoD mediate transcriptional activation of genes in repressive chromatin: a mechanism for lineage determination in myogenesis. Genes Dev 1997;11:436–450.
95 McKinnell IW, Ishibashi J, Le Grand F, Punch VG, Addicks GC, Greenblatt JF, Dilworth FJ, Rudnicki MA: Pax7 activates myogenic genes by recruitment of a histone methyltransferase complex. Nat Cell Biol 2008;10:77–84.
96 Kawabe Y, Wang YX, McKinnell IW, Bedford MT, Rudnicki MA: Carm1 regulates Pax7 transcriptional activity through MLL1/2 recruitment during asymmetric satellite stem cell divisions. Cell Stem Cell 2012;11:333–345.
97 Bergmann O, Bhardwaj RD, Bernard S, Zdunek S, Barnabe-Heider F, Walsh S, Zupicich J, Alkass K, Buchholz BA, Druid H, Jovinge S, Frisen J: Evidence for cardiomyocyte renewal in humans. Science 2009;324:98–102.
98 Kikuchi K, Poss KD: Cardiac regenerative capacity and mechanisms. Annu Rev Cell Dev Biol 2012;28:719–741.
99 Aurigemma GP: Diastolic heart failure – a common and lethal condition by any name. N Engl J Med 2006;355: 308–310.
100 Loffredo FS, Steinhauser ML, Jay SM, Gannon J, Pancoast JR, Yalamanchi P, Sinha M, Dall'Osso C, Khong D, Shadrach JL, Miller CM, Singer BS, Stewart A, Psychogios N, Gerszten RE, Hartigan AJ, Kim MJ, Serwold T, Wagers AJ, Lee RT: Growth differentiation factor 11 is a circulating factor that reverses age-related cardiac hypertrophy. Cell 2013;153:828–839.
101 Mourkioti F, Kustan J, Kraft P, Day JW, Zhao MM, Kost-Alimova M, Protopopov A, DePinho RA, Bernstein D, Meeker AK, Blau HM: Role of telomere dysfunction in cardiac failure in Duchenne muscular dystrophy. Nat Cell Biol 2013;15:895–904.
102 Zhang Y, Li TS, Lee ST, Wawrowsky KA, Cheng K, Galang G, Malliaras K, Abraham MR, Wang C, Marban E: Dedifferentiation and proliferation of mammalian cardiomyocytes. PLoS One 2010;5:e12559.
103 Kubin T, Poling J, Kostin S, Gajawada P, Hein S, Rees W, Wietelmann A, Tanaka M, Lorchner H, Schimanski S, Szibor M, Warnecke H, Braun T: Oncostatin M is a major mediator of cardiomyocyte dedifferentiation and remodeling. Cell Stem Cell 2011;9: 420–432.
104 Gearing DP, Comeau MR, Friend DJ, Gimpel SD, Thut CJ, McGourty J, Brasher KK, King JA, Gillis S, Mosley B, et al: The IL-6 signal transducer, gp130: an oncostatin M receptor and affinity converter for the LIF receptor. Science 1992;255:1434–1437.
105 Mozaffarian A, Brewer AW, Trueblood ES, Luzina IG, Todd NW, Atamas SP, Arnett HA: Mechanisms of oncostatin M-induced pulmonary inflammation and fibrosis. J Immunol 2008;181:7243–7253.
106 Tumpel S, Rudolph KL: The role of telomere shortening in somatic stem cells and tissue aging: lessons from telomerase model systems. Ann NY Acad Sci 2012;1266:28–39.
107 Leri A, Franco S, Zacheo A, Barlucchi L, Chimenti S, Limana F, Nadal-Ginard B, Kajstura J, Anversa P, Blasco MA: Ablation of telomerase and telomere loss leads to cardiac dilatation and heart failure associated with p53 upregulation. EMBO J 2003;22:131–139.
108 von Zglinicki T: Oxidative stress shortens telomeres. Trends Biochem Sci 2002;27:339–344.
109 Sahin E, Colla S, Liesa M, Moslehi J, Muller FL, Guo M, Cooper M, Kotton D, Fabian AJ, Walkey C, Maser RS, Tonon G, Foerster F, Xiong R, Wang YA, Shukla SA, Jaskelioff M, Martin ES, Heffernan TP, Protopopov A, Ivanova E, Mahoney JE, Kost-Alimova M, Perry SR, Bronson R, Liao R, Mulligan R, Shirihai OS, Chin L, DePinho RA: Telomere dysfunction induces metabolic and mitochondrial compromise. Nature 2011;470:359–365.

Stefan Tümpel, PhD
Leibniz Institute for Age Research, Fritz Lipmann Institute e.V. (FLI)
Beutenbergstrasse 11
DE–07745 Jena (Germany)
E-Mail tuempel@fli-leibniz.de

Rudolph KL (ed): Adult Stem Cells in Aging, Diseases and Cancer.
Else Kröner-Fresenius Symp. Basel, Karger, 2015, vol 5, pp 40–59 (DOI: 10.1159/000366570)

Mechanism of Functional Alterations in Hematopoietic Stem Cell Aging

Yohei Morita

Leibniz Institute for Age Research, Fritz Lipmann Institute, Jena, Germany

Abstract

Aging leads to functional decline of hematopoietic stem cells (HSCs), including alterations of self-renewal and differentiation, which is thought to contribute to age-related hematopoietic disorders, defects in immune functions, and an increase in hematopoietic malignancies. Different mechanisms could influence the functional decline or the malignant transformation of HSCs during aging. (a) The maintenance of HSC quiescence: low cell cycle activity is a characteristic sign of HSCs, and this may prevent the accumulation of replication-induced DNA damage and epigenetic alteration of the DNA. However, quiescent HSCs are also more vulnerable to accumulate DNA damage as homologous recombination-dependent DNA repair is only active in cycling cells. (b) DNA damage checkpoints: apoptosis and senescence could prevent the survival of damaged and/or mutant HSCs but may simultaneously provoke aging by diminishing the self-renewal of aging HSCs. Clarification of molecular changes in aging HSCs will not only help our understanding of age-related bone marrow failure and immunosenescence, but will also provide insights into the molecular basis for increased hematopoietic malignancies during aging. An understanding of these age-dependent processes could ultimately provide a rational basis for the development of new treatment and prevention strategies for age-associated bone marrow failure, immune dysfunction, and leukemia development.

Characteristics of Hematopoietic Stem Cells

Hematopoietic stem cells (HSCs) are the best-characterized somatic stem cells in the mammalian system [1] ensuring the maintenance and the development of all blood components, providing about 10^{11}–10^{12} new blood cells daily in humans [2]. HSCs can self-renew and produce progenitor cells differentiating into mature cells, thus fitting the concept of hierarchal organization. A hierarchical structure in the hematopoietic system is experimentally documented in a way that the most primitive stem cells reside at the apex of the hierarchy capable of both, self-renewal and generation of progenitor cells differentiating into a variety of mature cell types (fig. 1) [3].

The most primitive HSCs are long-term HSCs (LT-HSCs) that are characterized by a broad range of surface markers [4–10]. LT-HSCs are negative

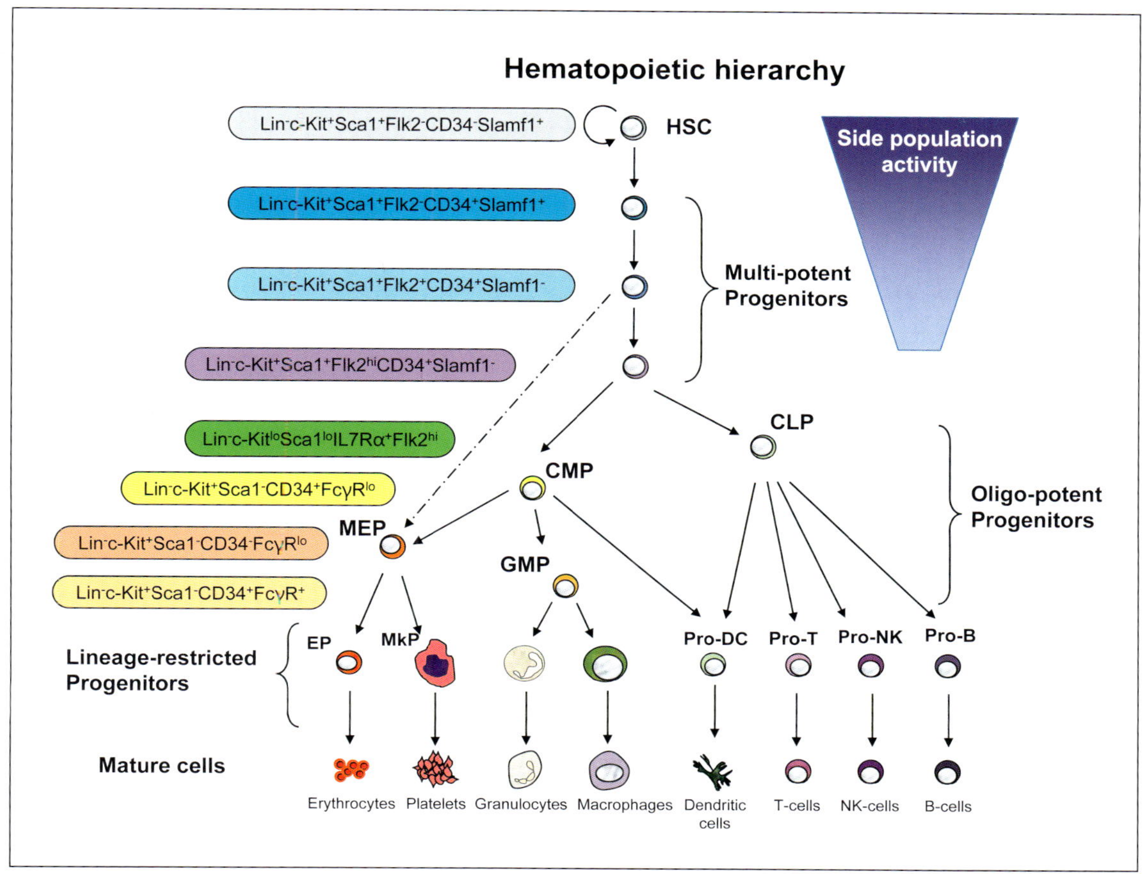

Fig. 1. Overview of the hematopoietic hierarchy and cell surface markers characterizing the different progenitor cells [18]. HSCs have self-renewal activity and first differentiate into multipotent progenitors that produce CMPs and CLPs. CMPs can differentiate into granulocyte/macrophage progenitors (GMPs) and megakaryocyte/erythrocyte progenitors (MEPs). CMPs and CLPs are responsible for myelopoiesis and lymphopoiesis, respectively.

for lineage markers, CD48, CD244, CD34, Flt3, and N-cadherin, but positive for Tie2, Endoglin, Thy1 (preferably expressed on lower levels), Sca-1, c-Kit, CD38 and CD150. In experimental setups, LT-HSCs are mainly sorted as $CD150^{+}CD34^{lo/-}$ $Flt3^{-}KSL$ cells (low or negative of CD34, Flt3-negative, lineage-negative, CD150-positive, Sca-1-positive and c-Kit-positive). This approach allows enrichment for LT-HSCs – when transplanted as single cells, 30–50% of the purified cells have the potential for long-term multi-lineage reconstitution in lethally irradiated mice by differentiating in all blood cell types [9]. The true percentage of LT-HSCs in this population might be underestimated due to experimental error in single cell transplants. In contrast to LT-HSCs, short-term HSCs (ST-HSCs) have already initiated differentiation and lack long-term reconstitution potential [11–15]. ST-HSCs differentiate into multipotent progenitor cells (MPPs) that lack self-renewal potential [8] but can differentiate into common myeloid and lymphoid progenitors (CMPs [16] and CLPs [17]), which further differentiate into functional blood cells.

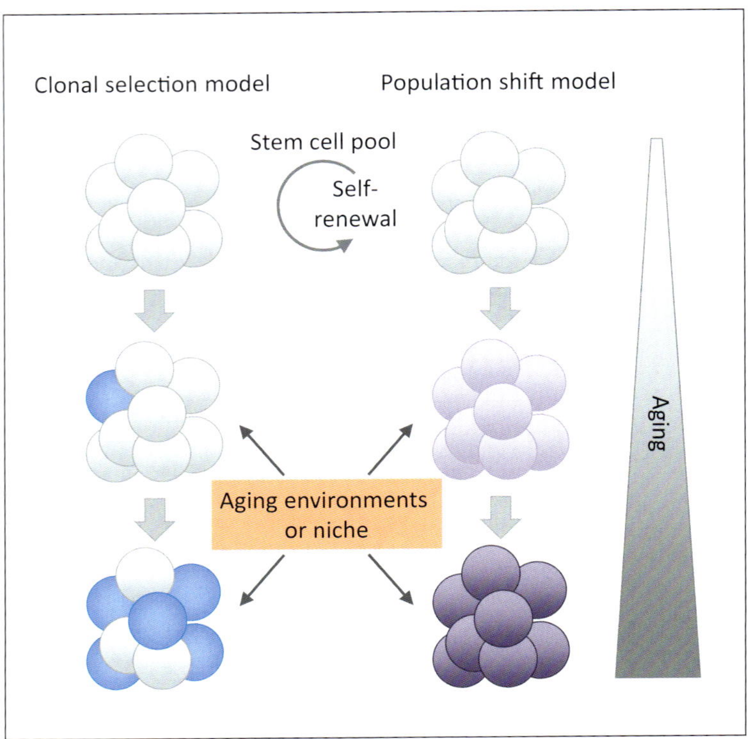

Fig. 2. Current models of stem cell aging [26]. The 'clonal selection model of stem cell aging' suggests that certain stem cell clones that bear distinct functional potentials are able to outcompete other types of clones during aging. Over time, such clones will come to predominate the stem cell pool, whereas others become diminished. In contrast, the 'population shift model of stem cell aging' suggests that the functional potential of all the cells in the stem cell population is equivalent at any stage of ontogeny, but the potential gradually and coordinately changes over time. These models are not mutually exclusive and both may be instructed by cell-intrinsic and non-cell autonomous cues originating from the aging micro- or systemic environment.

Functional Alteration of Hematopoietic Stem Cells during Aging

Previous investigations indicated that clonal selection is contributing to the process of HSC aging (fig. 2). Work from Christa Muller-Sieburg first showed that there are three different populations of HSCs: (a) balanced HSCs (Bala), (b) lymphoid-biased HSCs (Ly-bi), and (c) myeloid-biased HSCs (My-bi). All types of HSC produce myeloid and lymphoid cells in blood; however, Ly-bi HSCs predominantly generate lymphoid cells, and My-bi HSCs predominantly generate myeloid cells compared with Bala HSCs producing both lineages. It was shown that the different HSC subtypes show no obvious aging phenotype, but the relative proportion of these individual subpopulations changes during aging. This population shift leads to differences in the reconstitution capacities of the total pool of HSCs during aging. The My-bi HSCs accumulate with aging, whereas the other 2 subpopulations of HSCs decrease, resulting in an age-dependent skewing of hematolymphopoiesis with a decrease in lymphopoiesis and an increase in myelopoiesis. This phenotype is mouse strain-independent only showing minor inter-strain variations, e.g. a lower percentage of My-bi HSCs in DBA mice compared to C57BL/6 mice. In summary, the selective survival of My-bi HSCs appears to be an important contributor to impaired lymphocyte production in aging mice [19, 20]. Similar findings were recapitulated by other laboratories [21]. Recent studies showed that the CD150 marker can distinguish myeloid-biased from lymphoid-biased HSCs. Transplantation of $CD150^{high}CD34^{lo/-}$KSL cells results in myeloid-dominant reconstitution, whereas

$CD150^{low}CD34^{lo/-}$KSL cells show a balanced lymph-myeloid reconstitution [22, 23]. $CD150^{high}CD34^{lo/-}$KSL cells have a molecular profile that appears myelo-erythroid primed, while $CD150^{low}CD34^{lo/-}$KSL cells are more lymphoid primed. The greater self-renewal potential of myeloid-biased $CD150^{high}CD34^{lo/-}$KSL cells resulted in their expansion in the primitive stem cell pool during aging. Together, these studies support the model of clonal selection of subpopulations of HSCs during aging (fig. 2). However, the model remains under debate; an important question is to delineate the plasticity of different subpopulations of HSCs and whether one subpopulation can give rise to another subpopulation of HSCs.

Due to the phenotypic and quantitative changes of the stem cell pool in the hematopoietic system during aging, it is of great interest to estimate the stem cell pool size. Dr. de Haan introduced a retroviral cellular barcoding tool to track the clonality of HSCs [24]. Retroviral barcoded vector is used for transduction of HSCs in vitro, assuming that only one virus particle infects one HSC. After transplantation of single-infected HSCs, deep sequencing analysis can be performed on the different output populations (B cells, T cells and granulocytes) at different time points. Using a variety of barcodes, it can be estimated to what quantity and quality certain HSC subclones contribute to reconstitution. This system provides an intelligent tool to track HSCs during reconstitution and estimate their cell destiny. Using this system, Dr. de Haan showed that contribution of HSCs to mature blood cells in peripheral blood stabilizes at 12 weeks after transplantation. The clonal composition within each lineage after the stabilization showed good correlation, but the output of individual HSCs in HSC population of bone marrow (BM), and granulocytes, T cells and B cells of peripheral blood was quite different. To investigate differences of clonality between young and old HSCs, they transplanted young and old mixed HSCs (1:2 ratio) transduced with barcoded virus into lethally irradiated mice. They showed that old HSCs cannot efficiently produce mature cells in peripheral blood compared to young HSCs, although old HSCs can stay in the HSC compartment of BM, indicating diminished differentiation activity of old HSCs. They also showed that the clonality of old HSCs is always higher than that of young HSCs, whereas clone size of old HSCs is always smaller in the recipients. These data may imply that homing efficiency of old HSCs is equivalent to young HSCs and that old HSCs need to be proliferated more to maintain hematopoiesis because of the diminished repopulating activity.

Dr. Rodewald investigated the contribution of HSCs to hematopoiesis in steady state using a different cell tracing system. To exclude possible effects on hematopoiesis by viral insertion or transplantation, Dr. Katrin Busch in the Rodewald laboratory established a mouse line enabling in vivo lineage tracing. The system uses tamoxifen-inducible YFP expression in HSCs. In response to tamoxifen injection, a small number of functional HSCs were labeled by YFP, which was confirmed by measuring repopulating activity of single YFP+ HSCs after transplantation. The labeled HSCs differentiate into progenitors and eventually produce mature cells with YFP labeling. They analyzed the relationship between HSC and progenitors by YFP labeling frequency in each hematopoietic population from approximately 100 mice, and found strong correlations between LT-HSC and ST-HSC, ST-HSC and MPP, and MPP and CMP. In contrast, no correlation between MPP and CLP was found. These data suggest that conventional differentiation pathway from LT-HSCs to granulocytes is solid, but that CLPs might be not directly generated from MPPs. They also calculated based on limiting dilution analysis how many HSCs contribute to B cell, T cell and granulocyte progenitors, and estimated that of a total of 17,000 HSCs in a whole body, over time 5,000 HSCs are productive for the generation of mature blood cells, and 60, 110 and 700 HSCs contribute to T cell, B cell and granulocyte production, respectively, at single time points.

To analyze differentiation pathways downstream of MPPs, Dr. Rodewald's team utilized a Cre-driven double labeling system that can mark HSCs with YFP or RFP in the same mice. The ratio of YFP and RFP can be analyzed in each progenitor population to estimate the differentiation pathways. In this analysis, they again showed a correlation between MPP and CMP, and less correlation between MPP and CLP. They also investigated the branching point of B and T cells, and showed poor correlation between B and T cell production, indicating that T and B cells are not differentiated from a common pathway. Together, they concluded that hematopoiesis in steady state is a highly polyclonal process and that the pathways from HCS to myeloid but not lymphoid cells are continuous.

This newly described experimental system may also help to delineate molecular mechanism leading to stem cell aging or the selection of subpopulations of stem cells during aging. Previous studies used the technique of single cell transplantation to characterize HSC aging. Taking advantage of the known HSC markers (see above), HSCs were prospectively isolated and purified from C57BL/6 mice for evaluating the impact of aging on the stem cell compartment using single-cell repopulation experiments and limiting dilution analysis [24, 25]. These studies suggested that aging was accompanied by a steady increase in the frequencies of primitive hematopoietic cells, but the repopulating capacity of aged HSCs was reduced on a per cell basis. These observations again support the clonal selection model, where certain stem clones bearing distinct functional potentials are able to outcompete other types of clones during aging (fig. 2). Over time, such clones will predominate the stem cell pool, whereas others become diminished.

Molecular Mechanisms in Hematopoietic Stem Cell Aging

As mentioned above, self-renewal activity, differentiation potential and pool size of HSC are altered during aging. A recent study revealed that Batf, a transcription factor of the AP-1/ATF superfamily, is induced by DNA damage and contributes to the diminished self-renewal activity and myeloid skewing of HSCs by induction of HSC differentiation [27]. Interestingly, lymphoid-biased HSCs were more sensitive to DNA damage-induced, BATF-dependent depletion, suggesting that this checkpoint of DNA damage-induced stem cell differentiation may contribute to aging-associated impairments in lymphopoiesis.

During the Else Kröner-Fresenius Symposium, Dr. Geiger presented new experimental data that noncanonical Wnt signaling regulates HSC aging via induction of Cdc42 activity and cell polarity. Since it is thought that polarity is important for HSCs in symmetric/asymmetric divisions and interaction with niche cells [28, 29], Dr. Geiger and his colleagues analyzed cell polarity of young and old HSCs using immunostaining of polarity complexes including Cdc42, the Rho family GTPase, which is known as a regulator and a marker of cell polarity. Cdc42 functions as a molecular switch in signal transduction by shifting to GTP-bound or GDP-bound forms [30]. The GTP-bound form is the active form and can interact with effector proteins to regulate actin polymerization, integrin clustering, migration and adhesion. Dr. Geiger's group showed that 60% of young HSCs but only 20% of old HSCs exhibit a polarized Cdc42 localization. The group also investigated activity of Cdc42 in young and old HSCs comparing total Cdc42 with the active form Cdc42-GTP by pull-down assays. The experiments revealed that Cdc42 activity is significantly higher in old HSCs compared to young HSCs. They tested whether Cdc42 activity influences cell polarity and HSC aging using Casin, an inhibitor of Cdc42-GTP. The result showed that Casin treatment reverts the apolar state of aging HSCs to polar state, and improves myeloid skewing and cell number in old HSCs. These results indicate that cell polarity is important to preserve functionality of HSCs [31]. The question then arose: why and how is Cdc42 activity elevated upon aging? To answer this question, Dr. Geiger examined the expression level of Wnt family members in young and old HSCs, because it was known that Wnt signaling is involved in aging and Cdc42 activity in other tissue [32, 33]. They showed that noncanonical Wnt5a is expressed in old HSCs but not in young HSCs at both RNA and protein levels. In addition, downstream proteins of canonical Wnt signaling, β-catenin and Axin2, are downregulated in old HSC. These data indicated signaling shift from canonical Wnt to noncanonical Wnt signaling in HSCs during aging. The question is whether the modification of Wnt signaling influences functionality of aging HSCs. Dr. Geiger's group observed that Wnt5a treatment inhibits canonical Wnt signaling, and induced Cdc42 activation and apolar state in young HSCs. Conversely, knockdown of Wnt5a induced canonical Wnt signaling, Cdc42 inactivation and polar state of old HSCs and improved the aging phenotypes of HSCs in terms of myeloid skewing and expansion of pool size in transplantation assay. Together, Dr. Geiger concluded that shifting from canonical Wnt signaling to noncanonical Wnt signaling promote HSC aging via induction of Cdc42 activity (fig. 3). However, it remains unclear why noncanonical Wnt5a is elevated with age and how cell polarity maintains and renews HSCs.

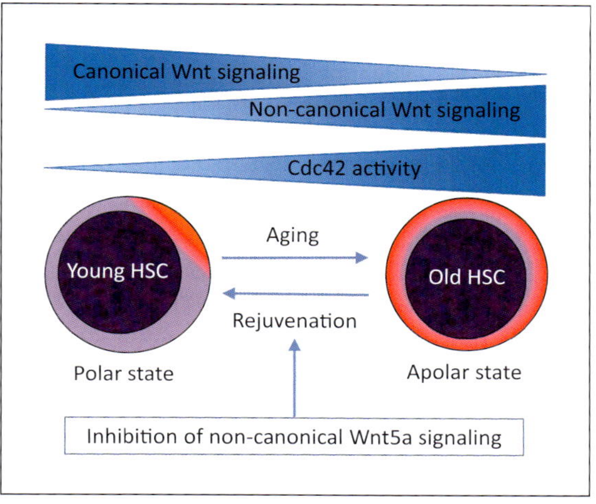

Fig. 3. Regulation of HSC aging by Wnt5a. Noncanonical Wnt signaling is elevated with age, while canonical Wnt signaling is reduced. Noncanonical Wnt5a induces Cdc42 inactivation and an apolar state in HSCs, resulting in HSC aging. Inhibition of Wnt5a rejuvenates old HSCs in terms of expanding HSC pool size and myeloid skewing.

Epigenetic Regulation of Stem Cell Function

Recently, the hypothesis of an epigenetic deregulation of gene expression in aging stem cells has emerged [34, 35]. According to this hypothesis, the loss of epigenetic control might contribute to stem cell aging. In detail, the loss or gain of the organization in heterochromatin or euchromatin, or relocalization of heterochromatin would result in differentially expressed transcripts, or complete random expression. Members of the polycomb (PcG) family are known to modulate epigenetic regulation of gene expression. PcGs were first identified as a conserved group of genes in *Drosophila*, required in multimeric protein complexes to maintain the expression of *Hox* genes [36–39]. PcG proteins interact with chromosomal elements, termed cellular memory modules. They have a repressive function, which is stable over generations and can only be overcome by germline differentiation processes.

Several mammalian homologues of *Drosophila* PcGs have been identified. PcG proteins were shown to form two complexes: PRC1 and PRC2. Both complexes are necessary for the repressive effect. PRC2 initiates PcG-mediated repression, whereas PRC1 maintains the repression.

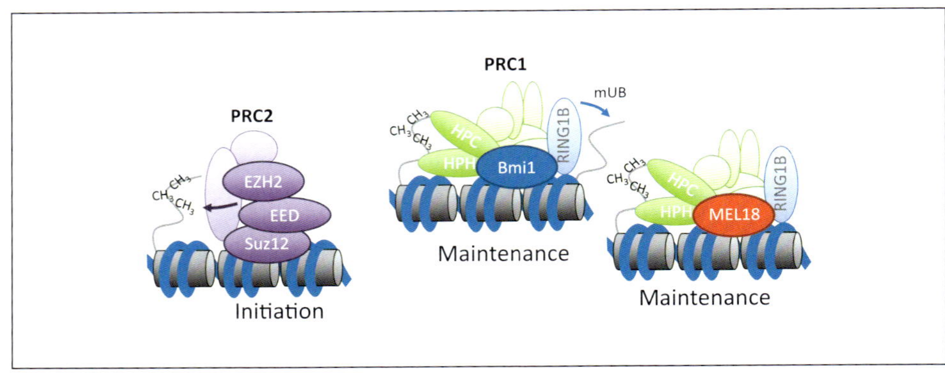

Fig. 4. Composition of the PRC1 and the PRC2 complex in the context of the initiation or maintenance of methylation marks. The composition of the PRC2 complex including EZH2, EED and Suz12 is essential for the initiation of methylation marks on the DNA. Alternative compositions form the PRC1 or PRC1b complex. Both protein complexes PRC1 and PRC1b are responsible for maintaining methylation marks on the DNA, e.g. during replication of DNA. PRC1 and PRC1b differ only in one component: PCR1 includes Bmi1, whereas PRC1b includes alternatively Mel18. Bmi1 and Mel18 are exchangeable in vitro but have diverse functions in the regulation of adult stem cells.

PRC2 includes the histone-methyltransferase Ezh2 setting trimethylmarks on lysine 27 of histone H3 tail. This repressive mark is recognized by the chromodomain proteins in the PRC1 complex [40, 41]. PRC1 is larger than PRC2 and has enzymatic activity as an E3 ubiquitin ligase. This is mediated by RING1B, which is forming a heterodimer with Bmi1 acting as a regulator and enhancer of the complex (fig. 4).

The mono-ubiquitin mark is set on histone H2A. This mark might be important for PcG repression, and it keeps the polymerase II stalled at the PcG targets. All chromatin marks are highly dynamic and can be removed by demethylases. Depending on the context, the marks are continuously set or removed [39, 42–50].

Bmi1 and Mel18 in the PRC1 complex were shown to be exchangeable. Because of 75% similarity, they display the exact same biochemically functions in vitro. The knockout (KO) phenotypes overlap but they are not exactly the same, e.g. Bmi1 and Mel18 display distinct functions in the regulation of adult stem cells. Bmi1 is essential in the maintenance and the regulation of self-renewal in HSCs [51–53]. Bmi1$^{-/-}$ mice have markedly reduced numbers of HSCs that exhibit severe defects in self-renewal and can only transiently contribute to hematopoiesis [54]. It was proven that Bmi1 alone is not the only regulator of self-renewal but a well-balanced expression of Bmi1 and Mel18 is required to maintain stem cell differentiation and self-renewal [55].

It is known that there is a profound failure of Bmi1-deficient HSC in reconstituting irradiated recipients in contrast to wild-type (WT) HSCs [54]. Moreover, Bmi1 was not only required for the self-renewal and reconstituting activity of normal HSCs but also required by leukemic stem cells [56]. Bmi1-deficient hematopoietic cells transformed by Hoxa9 and Meis1 (a homeobox protein; frequent coexpression of HoxA9 and Meis1 genes was found in infant acute lymphoblastic leukemia with MLL rearrangement) could form acute myeloid leukemia. However, the leukemic cells of Bmi1$^{-/-}$ HSCs were not transplantable, whereas HoxA9/Meis1-induced leukemia cells from Bmi1$^{+/+}$ HSCs were transplantable, indicating that Bmi1 has not only important function to maintain self-renewal of nontransformed HSCs but also the self-renewal of leukemic stem cells [56].

Bmi1 overexpression cooperates with phosphate-tensin homolog (PTEN) loss in promoting prostate carcinogenesis by depressing the Ink4a/Arf tumor suppressor pathway [57]. Germline deletion of Bmi1 in mice led to overall growth retardation, premature death, posterior transformation, neurological abnormalities, and severe hematopoietic defects [58]. Further investigations revealed that Bmi1 is required for the self-renewal and maintenance of both HSCs and neural stem cells [54, 59]. Dr. van Lohuizen found that Bmi1 controls the differentiation in the mammary epithelial stem cells and progenitor cells. Mammary tissue from Bmi1-deficient mice displayed a premature differentiation leading to a dramatic loss of repopulating capacity of mammary stem cells [60]. Loss of $p16^{Ink4a}/p19^{Arf}$ (also known as Cdkn2a) rescued the phenotype in mammary tissue transplantation experiments. Together with evidence from other stem cell compartments, the de-repression of the $p16^{Ink4a}/p19^{Arf}$ locus was suggested to be responsible, at least in part, for the abnormalities developed in Bmi1-deficient mice [61–63]. Although the deletion of $p16^{Ink4a}/p19^{Arf}$ in Bmi1-deficient mice improved stem cell function, the cellular composition of the BM microenvironment remained significantly altered in double-null mice [64]. In addition, the body weight and overall survival remained reduced compared to WT littermates.

The mentioned methylation of lysine 27 residue on histone H3 (Lys27H3) was shown to be important in HSCs. This led to the hypothesis that PcGs are able to regulate the heterogeneity of HSCs [53]. In response to the methylation of Lys27 on H3, secondary genes are repressed. The methylation mark can be copied during cell division and the repressed state is inherited to the daughter cells. But under the assumption that PcGs could be absent or less abundant in HSCs, the methylation mark of Lys27H3 would not be copied faithfully during cell division. Therefore, one of the daughter cells would activate the normally repressed genes, whereas the other daughter cell keeps the repressing mark. This unequal activation of genes that should be repressed results in two characteristically different daughter cells and increases the variability of HSCs in a 'not-programmed fashion'. Picturing this effect for the whole genome, this could lead to unregulated processes in all dividing cells. On the other hand, if PcGs are active, the methylation mark is copied during replication and keeps target genes switched off in all daughter cells. The PRC1 complex was already shown to display diverse activity in copying methylation marks in young and old cells due to altered recruitment of the complex to the silenced locus, resulting therefore in heterogeneous cell populations regarding the methylation state [65].

Gerald de Haan's group [66] also reported epigenetic influences on HSCs. HSCs can show divergent aging phenotypes: (a) HSCs can exhaust resulting in BM failure or (b) HSCs can become hyperproliferative leading to neoplasia and leukemia. If one would know more about the epigenetic regulation of HSCs, these adverse phenotypes might be avoided. Despite the fact that HSCs are able to self-renew and display distinct telomerase expression, they have a limited number of cell divisions lowering the quality of functional HSCs with increasing numbers of divisions. It was already shown that Bmi1, involved in the PRC1 complex, is a regulator of stem cell maintenance. PRC1 recognizes the methylation mark H3K27 and contains various Cbx proteins, which are directly binding to the methylation mark. One PRC1 complex molecule always contains only one specific Cbx member, suggesting exclusive functions for Cbx proteins. To identify Cbx activity in the hematopoietic system Dr. de Haan and colleagues overexpressed Cbx2, Cbx4, Cbx7 or Cbx8 in HSC-enriched BM cells by using retroviral infection. Even though Cbx2, Cbx4,

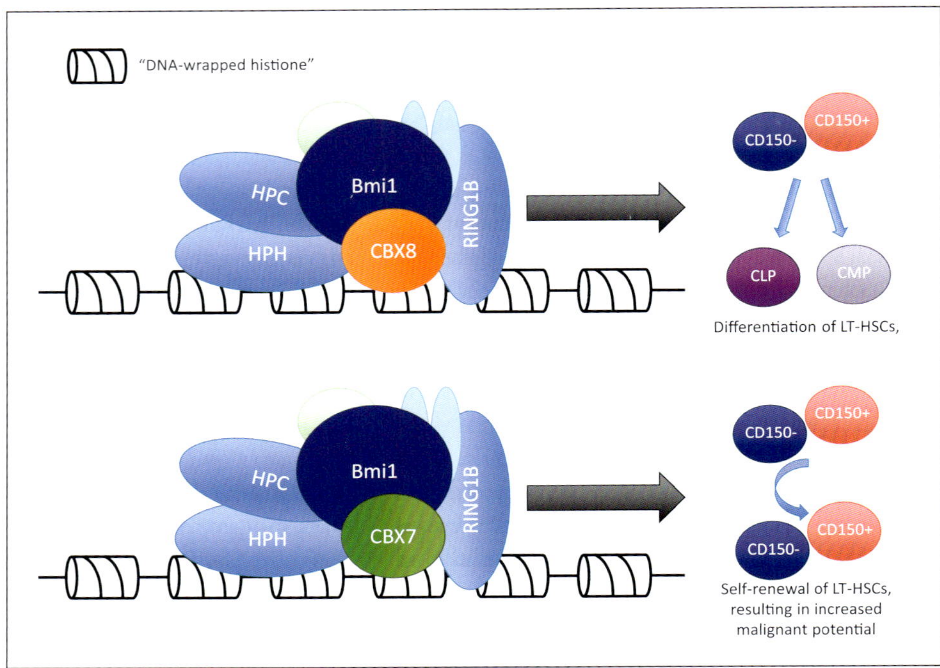

Fig. 5. Function of Cbx proteins in the PRC1 complex. Cbx7 and Cbx8 have differential roles in the PRC1 complex in epigenetic regulation of HSCs. Whereas Cbx8 combined with the PRC1 complex results in differentiation of HSCs, Cbx7, integrated in the PRC1 complex, leads to self-renewal of LT-HSCs.

Cbx7 and Cbx8 are homologues, they display different effects on self-renewal of HSCs. Overexpression of Cbx2, Cbx4 and Cbx8 in HSCs results in a loss of reconstitution potential in vivo. In contrast to the data, Cbx7 overexpression increased self-renewal of HSCs shown by enhanced in vivo repopulation capacity. However, most transplanted animals died due to leukemia, indicating that increased self-renewal of HSCs in response to Cbx7 overexpression resulted in cancer development.

In conclusion, it can be hypothesized that PRC1 complex containing Cbx2, Cbx4 or Cbx8 induces differentiation of HSCs, whereas PRC1 complex including Cbx7 induces self-renewal but might also initiate malignant transformation (fig. 5). These results are also in line with endogenous expression of Cbx7, which is restricted to the most primitive cells of the hematopoietic system, and Cbx8, which shows a more constant expression throughout hematopoiesis.

The counterplayers of PcGs are the Thritorax group proteins (Trx-G) that were shown to maintain the expression of Hox and other developmental genes by epigenetic modification of histones [67]. Trx-G and PcGs control the maintenance, but not the initiation of gene expression as epigenetic guardians of cell identity. Furthermore, there is also evidence for the involvement of Trx-G in cellular proliferation and tumorigenesis [68–70]. In recent studies, Fehling and colleagues found that Mll5, the latest MLL/Trx-G-member, impairs the function of primitive hematopoietic cells. Mll5 is expressed in most somatic tissues. To obtain a first understanding of potential physiological functions, Mll5-deficient mice were generated. Intercrosses of heterozygous Mll5$^{+/-}$ mice produced viable homozygous KO/KO offspring; however, approximately 40% of homozygous Mll5-deficient pubs died for unknown reasons during the first days of postnatal life. Mll5$^{-/-}$ pups surviving this critical period developed into outwardly healthy adults. Further analyses revealed a multitude of phenotypic abnormalities, including complete male sterility, impaired female fertility, retarded growth, and defective lymphopoiesis. In detail, B and T cell numbers were reduced in Mll5 KO mice, and mutant HSCs showed strong reconstitution defects in transplantation experiments, CD4/CD8 double-positive and double-negative (DN1 and DN2) cells in the thymus were reduced proportionally, while ETPs were almost absent. Additionally, the total number of hematopoietic stem/progenitor cell populations was decreased [71].

DNA methylation controls gene expression cooperatively with histone modification in an epigenetic manner. DNA methylation involves additional reaction of methyl groups to cytosine pyrimidine ring and suppresses gene expression [72]. The modification of DNA is inherited in cell division by DNA methyl transferase (Dnmt). There are 3 different Dnmts: Dnmt1, 3a and 3b. Dnmt1 exhibits potential to maintain methylation, which is necessary to preserve DNA methylation in every cell division. On the other hand, Dnmt3a and Dnmt3b are known as de novo DNA methyl transferases to add newly methyl group, which is especially important for genome imprinting in early development [73, 74].

Recently, Dr. Goodell and her colleagues have shown that Dnmt3a KO HSCs lose differentiation capacity rapidly upon serial transplantation, while HSC number dramatically increases, indicating that Dnmt3a has an important role in fate decision between differentiation and self-renewal [75]. How does Dnmt3a deletion limit differentiation and promote self-renewal? To address this question, they performed RNA sequencing in WT HSCs, WT B cells, Dnmt3a/b KO HSCs and Dnmt3a/b KO B cells, and analyzed the expression level of genes that are expressed higher in WT HSCs compared to WT B lymphocytes. This analysis revealed that Dnmt3a/b KO HSCs express the genes slightly higher than WT HSCs. Interestingly, B lymphocytes from Dnmt3a/b KO mice expressed the HSC-related genes at a higher level. These data indicate that Dnmt3a inhibits HSC-specific genes by DNA methylation, which allows proper differentiation of HSCs. Dr. Goodell focused on DNA methylation in HSCs to analyze more deeply the Dnmt3a-regulating pathway. First, the group performed reduced representative bisulfite sequencing (RRBS) that enables to analyze methylation status of most CpGs. However, they recognized the need of whole-genome analysis, because RRBS showed existing, both expected hypomethylated and unexpected hypermethylated, regions in the genome of Dnmt3a KO HSCs compared to WT HSCs, and the data had poor correlation with the gene expression analysis. Then, whole-genome bisulfite sequencing was performed in WT and Dnmt3a KO HSCs. In the process of the analysis, Dr. Goodell and her colleagues discovered conserved large unmethylated regions (>3.5 kb) called 'hypo-methylated canyons' in the genome of WT HSCs, which are larger than CpG island and especially located in gene loci encoding transcription factors [76]. Surprisingly, when they looked at top 12 large hypomethylated canyons, most of the genes located in these canyons were not expressed in HSCs. To address this unanticipated result, Dr. Goodell's group investigated chromatin methylation status of the canyons and found that hypomethylated canyons, containing genes that were not expressed in HSCs, were covered with H3K27me3 marks repressing gene expression. In contrast, canyons containing genes that were expressed in HSCs were covered with H3K4me3 marks activating transcription. Together, these data indicated that the types of methylation marks are critical to predict the expression of genes located in hypo-

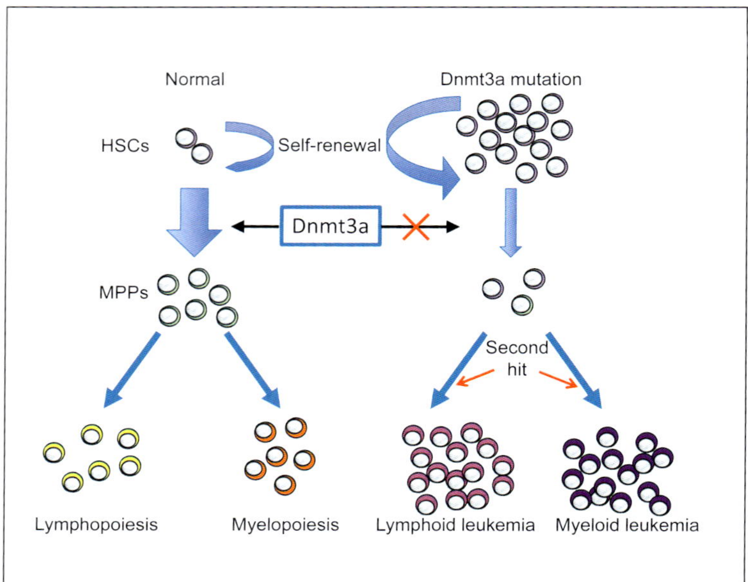

Fig. 6. Model of leukemogenesis by Dnmt3a mutation. Dnmt3a mutation leads to accumulation of HSCs and diminished HSC differentiation activity. Dnmt3a-mutated differentiated cells possess a stem cell program and evolve to varied types of leukemic cells by second hits.

methylated canyons of HSCs from WT mice. Next, Dr. Goodell investigated whether Dnmt3a deletion would affect hypomethylated canyons in HSCs. Generally, DNA methylation status should be maintained by Dnmt1. Recent studies revealed that Tet proteins, members of DNA hydroxylases, convert methylated cytosine to hydroxymethylated cytosine (HMC) that cannot be recognized by Dnmt1 resulting in loss of DNA methylation during DNA replication [77]. Dr. Goodell showed that the edges of hypomethylated canyons exhibit an erosion of DNA methylation in response to Dnmt3a deletion, and the edges are strongly marked with HMC, suggesting that Dnmt3a and Tet proteins cooperate to regulate gene expression in HSCs.

Interestingly, genes associated with human leukemia were enriched in hypomethylated canyons found in HSCs. Dr. Goodell presented data showing that Dnmt3a mutation accelerates leukemogenesis. It is known that Dnmt3 mutation is associated with hematologic disease in humans [78, 79]. The Goodell group identified overexpressed FLT3-ITD as a second mutation in Dnmt3a KO HSCs. Dnmt3a deletion accelerated hematologic disease and FLT3-ITD-overexpressed Dnmt3a KO HSCs developed T cell leukemia. Additionally, abnormal myeloproliferation in FLT-ITD knockin Dnmt3a heterozygous KO mice was shown. The data indicate that different mutations of Dnmt3a, homozygous or heterozygous, initiate different types of leukemia. Dr. Goodell summarized that Dnmt3a mutation expands the HSC number and maintains the stemness of the cells during differentiation, which causes various types of leukemia with second hit (fig. 6).

Recently, the Goodell group also performed the above-described epigenome analyses including histone modification analysis and transcriptome analysis in young and old HSCs, and showed that alterations in transcriptome and epigenome during aging associate with age-related changes in phenotypic and functional HSCs. Aging-associated epigenetic alteration may also contribute to age-related diseases such as leukemia [80].

DNA Damage Signaling in the Regulation of Hematopoietic Stem Cells

HSCs display decreased functionality in serial transplantation experiments. After 4–6 rounds of transplantation, reconstitution capability is strongly decreasing either due to a strong dilution or exhaustion of LT-HSCs. Previous study showed that single HSC can generate 300–1,000 HSCs in vivo. But repopulating activity of in vivo expanded HSC is reduced to 1/10–1/50 of that of freshly isolated, unexpanded HSCs. The difference in the repopulating activity of in vivo expanded HSCs showed high cell-to-cell variation. To investigate what causes these strong variations, Dr. Nakauchi and colleagues performed telomere length and telomerase activity assays in hematopoietic cells. They could prove moderate telomerase expression in all stem and progenitor cell compartments. The telomere length studies of donor-derived WT and $TERT^{-/-}$ (telomerase deficient) HSCs several months after transplantation displayed a clear shortening, even more pronounced in late generations of $TERT^{-/-}$ (G2 and G3) but also visible in transplanted WT HSCs. When LT-HSCs were injected in doses of 10 or 100 cells per recipient, shortening of telomeres was reduced compared with single-cell transplantations. Together, these data support the model that replicative stress and telomerase deficiency accelerate telomere shortening of LT-HSCs. These data suggest that BM transplantation should always include the maximum amount of HSCs to keep the replicative stress for the cells as low as possible.

Telomere shortening can contribute to DNA damage accumulation during aging. Despite the fact that most stem cells express moderate levels of telomerase, telomeres shorten during aging, which may contribute to impaired stem cell function and loss of organ regeneration during aging. Studies of mouse strains bearing mutations in DNA damage response or repair pathways strengthened the concept that DNA damage can impair the functional reserve of stem cell compartments [81–84]. Dr. Rossi reported that DNA damage accumulates in HSCs of aging mice and this accumulation of DNA damage appears to be very restricted to the stem cell compartment. Global gene expression profiling revealed that HSCs have an attenuated DNA repair activity compared to downstream progenitors. DNA damage responses including DNA repair pathways are downregulated in quiescent HSCs but not in cycling HSCs. HSCs cycle rarely and show little DNA repair activity, but damaged stem cells can repair DNA damage when they reenter the cell cycle [84]. Dr. Rossi's studies provided experimental evidence that HSCs that harbor unrepaired DNA damage can get depleted by apoptosis, and possibly in rare circumstances, by senescence. The epigenetic de-repression of the INK4a/ARF locus could be involved in the induction of stem cell senescence (see above). Deletion of $p16^{Ink4a}$ was shown to improve functional capacity of neural, pancreatic and HSCs in aging mice [85–87]. In addition to DNA damage and de-repression of $p16^{INK4a}$, telomere shortening represents another mechanism mediating cellular senescence, growth arrest and/or cell death during aging [88, 89].

Telomere dysfunction represents a specific form of DNA damage. In the absence of functional telomerase, telomere attrition both attenuates stem cell function and contributes to organismal aging in mice and humans. Late-generation telomerase-deficient mice with dysfunctional telomeres have a reduced lifespan and experience progressive degeneration in proliferative tissues such as skin, intestine, and blood [90–92]. The impact of telomere dysfunction on stem cell compartments in telomere dysfunctional mice was shown to depend on both cell intrinsic checkpoints and cell extrinsic alterations [93–95]. When telomeres become critically short, p53-dependent checkpoints are activated leading to senescence or apoptosis. The CDK inhibitor p21 is one of the downstream effectors of p53 that is responsible for cell cycle arrest. Deletion of p21 improves the maintenance and function of telomere dysfunctional stem cells, and extends the lifespan of late-generation telomerase-deficient mice [93]. However, p53-dependent checkpoints were also shown to have a beneficial effect on tissue homeostasis in the context of telomere dysfunction. Specifically, it was shown that p53-dependent responses mediate the clearance of chromosomal instable stem cells with critically short telomeres [96]. Follow-up studies revealed that selective inhibition of p53-dependent apoptosis or cell cycle arrest can improve stem cell maintenance and tissue function in aging telomere dysfunctional mice, but the combined abrogation of both checkpoint branches results in chromosomal instability at stem cell level and accelerated tissue aging [97].

Dr. de Haan also investigated the role of the cyclin-dependent kinase inhibitor p21 in the maintenance of functional HSC in aging telomerase WT mice with long telomere reserves. Deletion of p21 in mice of the 129/SvEv background resulted in an increase in HSC cycling and in the total number of HSCs in young mice but the loss of p21 was also associated with premature stem cell exhaustion in serial transplantation experiments and with increased chemosensitivity of HSCs in response to 5-FU treatment [98, 99]. In contrast, the KO of p21 in the C57Bl/6J background showed only minimal effects on HSC maintenance [100]. In addition, work from Rudolph and colleagues revealed that the deletion of p21 can improve stem cell maintenance and function in telomere dysfunctional ($Terc^{-/-}$) C57Bl/6J mice (see above) [91, 92]. These data indicated that p21 can have both positive effects on HSC maintenance by mediating HSC quiescence, but can also exert negative effects on HSC self-renewal when DNA damage checkpoints are activated in response to telomere dysfunction [93].

A different example for mouse strain-specific changes in the expression of HSC-regulating genes is the Slit2 gene. The mammalian SLIT genes belong to a highly conserved family of axon guidance molecules. Recently, they were also shown to play a role in acute and chronic lymphoid leukemia [101, 102]. Slit2 is a cell type-specific regulated and the most primitive population of HSCs that expresses high levels of Slit2, whereas committed progenitor cells express significantly lower levels of the gene, similar to the expression pattern of latexin. Slit2 expression also shows differences in D2 compared to B6 mice, in which C57BL/6J mice show higher levels of Slit2 [103]. Moreover, a number of 182 genes were strain-dependent regulated in HSCs of D2 compared to B6 mice. The question arises whether there are common specific transcription factors that regulate these genes. In this context, it could be important that gene expression in differentiated cells is quite often *trans*-regulated, whereas gene expression in stem cells (including HSCs) is rather *cis*-regulated [104, 105]. In addition, it remains a critical question, which of the strain-specific differences in gene expression that influence HSC function, self-renewal and aging is important for HSC aging in humans.

Genes Controlling Quiescence and Self-Renewal of Hematopoietic Stem Cells

Studies on a wide range of genetically manipulated model systems demonstrated that intrinsic HSC alterations can lead to the dysfunction of stem cells. The PTEN, a negative regulator of the PI3K-Akt pathway and tumor suppressor, is crucial for the inhibition of leukemogenesis. If PTEN is lost, ST-HSCs are strongly activated and LT-HSCs are depleted over time. PTEN-deficient HSCs engraft with normal efficiency after transplantation but hematopoietic reconstitution cannot be sustained due to a deregulation of the cell cycle. Loss of PTEN leads to myeloproliferative disease and transplantable leukemia since PTEN functions to restrict the proliferation of hematopoietic progenitor cells [106, 107].

Related to the function of PTEN is the cellular oncogene c-Myc. Loss of c-Myc leads to accumulation of LT-HSCs in the niche due to upregulation of adhesion molecules, especially N-cadherin. It was shown that this increased adhesion impaired the exit of LT-HSCs from the niche. Conversely, overexpression of c-Myc leads to the loss of HSCs due to premature differentiation associated with a downregulation of adhesion molecules. In summary, c-Myc is an important player in regulating the fate decision between self-renewal and differentiation [108].

As discussed above, cyclin-dependent kinase inhibitors also have a function in the regulation of stem cell quiescence and stem cell maintenance. The individual KOs for p18 and p21 were both shown to increase proliferation of HSCs. The deletion of p18 results in an increase of the functional HSC pool [109], whereas p21 deletion can lead to an early exhaustion of LT-HSCs in certain genetic backgrounds (see above). Together, it appears that stem cell quiescence strongly influences the maintenance of functional stem cells. Along these lines, it was recently shown that replication stress leads to functional exhaustions of HSCs in serial transplantation experiments by inducing alterations in DNA methylation [110]. Interestingly, p18 deletion rescued the effects of both serial transplantation and p21 deletion on the exhaustion of HSC self-renewal [111]. Whether these positive effects of p18 deletion on HSC self-renewal have anything to do with DNA methylation remains to be investigated.

Two other important intrinsic HSC regulators are MEF/ELF4 and early growth response 1 (Egr1). The myeloid elf-1-like factor (MEF/ELF4) regulates the quiescence of LT-HSCs. Deletion of MEF leads to enhanced quiescence of HSCs. The reconstitution potential of MEF-deleted HSCs is normal and the mice show an enhanced protection against myelotoxic stress and radiation [112, 113]. Egr1, as an immediate early response transcription factor and zinc-finger protein, is expressed in LT-HSCs. $Egr1^{-/-}$ mice show increased proliferation of HSCs resulting in the mobilization into the blood by so far unknown mechanisms, suggesting that Egr1 acts as a retention and quiescence factor in HSCs [114].

Trumpp and colleagues have shown that the $CD34^{lo/-}$KSL HSCs are subdivided into two populations. One is an active population which enters the cell cycle and is able to reconstitute mice but prematurely exhausts in serial transplantations. In addition, the BM contains a subpopulation of quiescent $CD34^{lo/-}$ HSCs that very rarely enter the cell cycle. These 'dormant' HSCs (d-HSCs) cycle only five times during the lifespan of a mouse and they are mostly responsible for injury repair but not for the daily regeneration of the blood system [115]. These long-term quiescent HSCs have a reduced metabolism and silenced replication machinery. d-HSCs become activated by injury signals, e.g. like 5-FU treatment, and can go back into dormancy after the successful replenishment of the hematopoietic compartment [115]. A recent study also showed that quiescent state in tissue stem cells including HSCs is comprised of two functionally distinct phases, G0 and alert phases. The alert stem cells induced by an injury have rapid regenerative activity. Interestingly, an injury induces the transition from G0 phase to alert phase in several different tissue stem cells. The systematic transition mechanism functions as an adoptive response to regenerate tissue in future injury [116].

An interesting task of HSC research is now to define the molecular pathways controlling the quiescence of HSCs. There are already some data about the importance of lipid rafts in quiescent HSCs. The PI3K-Akt-FoxO signaling pathway is inactive in freshly isolated HSCs but active in cycling progenitor cells. Thereby, the lipid raft sta-

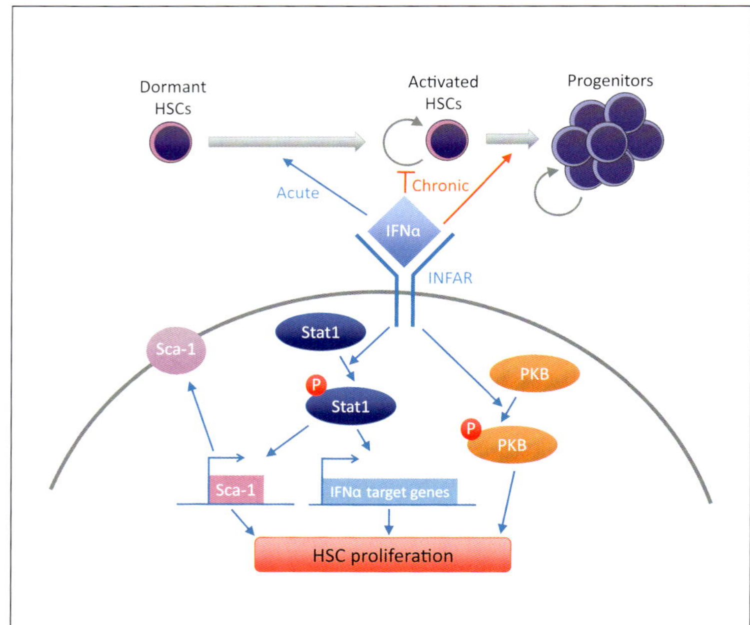

Fig. 7. Overview of the activating effects of IFNα on ST- and LT-HSCs. Acute treatment of mice with IFNα leads to an activation of dormant HSCs via activation of STAT1 and secondary upregulation of Sca1 expression resulting in proliferation of HSCs.

tus finely tunes cytokine signal levels and regulates Akt activity. Lipid rafts are cholesterol- and glycosphingolipid-enriched spots in the membrane of the cells containing a variety of functional molecules and acting as platforms for cellular functions. The size of the lipid rafts determines the signal intensity and functional outcome by formation of clusters. These clusters are absent in quiescent HSCs but can be clearly observed in activated stem cells [117]. The clustering of lipid rafts, in turn, was shown to be negatively regulated by TGF-β. TGF-β significantly inhibits cytokine-induced activation of c-Src, one of the lipid raft components, in HSCs. Although the precise molecular mechanism for this inhibition is so far unknown, Src family kinases could be key targets for TGF-β in affecting lipid raft reorganization [118].

Work from the Trumpp laboratory revealed that treatment with interferon-α (IFNα) increases HSC numbers. Interferons, especially IFNα, inhibit virus replication and are strongly produced during viral infection [119–122]. IFNα can directly activate HSCs. This signaling is not mediated by the BM niche but directly activates the dormant HSCs. Trumpp and colleagues showed that the stimulatory effects are mediated via the Jak/STAT signaling pathway. In detail, STAT1 is the mediator for the action of IFNα on d-HSCs. STAT1 activates Sca1 expression culminating in the proliferation of HSCs. Together, the results support a model indicating that acute interferon treatment activates dormant HSCs and a chronic interferon treatment leads to the exhaustion of the HSC pool (fig. 7).

The studies by the Trumpp lab also showed that IFNα activates dormant HSCs, which in turn are sensitized to killing by 5-FU treatment. IFNα treatment may also activate quiescent CML stem cells that could then be eliminated by current anticancer therapies such as imatinib. If the model holds true, interferon treatment may have the potential to improve treatment results in CML patients [123].

Conclusions

The above-described studies and meeting presentations suggest that there are age-dependent limitations in the maintenance of functional HSCs. Several molecular mechanisms have evolved to limit the cell cycle activity of HSCs, thereby preventing the exhaustion of HSC self-renewal occurring in response to replication stress. It will be interesting to study molecular mechanisms of quiescence control of HSCs in greater detail and to delineate the processes that limit HSC self-renewal in response to cell cycle activation. An emerging topic is to characterize epigenetic alterations that are induced by cell cycle activity. To improve healthy aging of the hematopoietic system, it will also be important to analyze how these mechanisms influence the initiation of hematopoietic malignancies at the cell-intrinsic level but also the selection of premalignant and fully malignant, mutant HSCs during cancer initiation and progression. The increasing understanding of molecular changes in aging HSCs will help to develop molecular therapies aiming to improve hematopoiesis in the elderly. In addition, the understanding of HSC aging will contribute to discover prevention strategies and new therapies for the treatment of hematolymphopoietic malignancies that show an increasing incidence in older people.

References

1 Ema H, Morita Y, Yamazaki S, Matsubara A, Seita J, Tadokoro Y, et al: Adult mouse hematopoietic stem cells: purification and single-cell assays. Nat Protoc 2006;1:2979–2987.

2 Mackey MC: Cell kinetic status of haematopoietic stem cells. Cell Prolif 2001; 34:71–83.

3 Bryder D, Rossi DJ, Weissman IL: Hematopoietic stem cells – the paradigmatic tissue-specific stem cell. Am J Pathol 2006;169:338–346.

4 Baum CM, Weissman IL, Tsukamoto AS, Buckle AM, Peault B: Isolation of a candidate human hematopoietic stem-cell population. Proc Natl Acad Sci USA 1992;89:2804–2808.

5 Goodell MA, Brose K, Paradis G, Conner AS, Mulligan RC: Isolation and functional properties of murine hematopoietic stem cells that are replicating in vivo. J Exp Med 1996;183:1797–1806.

6 Goodell MA, Rosenzweig M, Kim H, Marks DF, DeMaria M, Paradis G, et al: Dye efflux studies suggest that hematopoietic stem cells expressing low or undetectable levels of CD34 antigen exist in multiple species. Nat Med 1997; 3:1337–1345.

7 Kiel MJ, Yilmaz OH Iwashita T, Terhorst C, Morrison SJ: SLAM family receptors distinguish hematopoietic stem and progenitor cells and reveal endothelial niches for stem cells. Cell 2005; 121:1109–1121.

8 Morrison SJ, Weissman IL: The long-term repopulating subset of hematopoietic stem cells is deterministic and isolatable by phenotype. Immunity 1994;1:661–673.

9 Osawa M, Hanada K, Hamada H, Nakauchi H: Long-term lymphohematopoietic reconstitution by a single CD34-low/negative hematopoietic stem cell. Science 1996;273:242–245.

10 Spangrude GJ, Heimfeld S, Weissman IL: Purification and characterization of mouse hematopoietic stem cells. Science 1988;241:58–62.

11 Akashi K, He X, Chen J, Iwasaki H, Niu C, Steenhard B, et al: Transcriptional accessibility for genes of multiple tissues and hematopoietic lineages is hierarchically controlled during early hematopoiesis. Blood 2003;101:383–389.

12 Lerner C, Harrison DE: 5-Fluorouracil spares hemopoietic stem cells responsible for long-term repopulation. Exp Hematol 1990;18:114–118.

13 Ramalho-Santos M, Yoon S, Matsuzaki Y, Mulligan RC, Melton DA: 'Stemness': transcriptional profiling of embryonic and adult stem cells. Science 2002;298: 597–600.

14 Uchida N, Dykstra B, Lyons KJ, Leung FY, Eaves CJ: Different in vivo repopulating activities of purified hematopoietic stem cells before and after being stimulated to divide in vitro with the same kinetics. Exp Hematol 2003;31: 1338–1347.

15 Venezia TA, Merchant AA, Ramos CA, Whitehouse NL, Young AS, Shaw CA, et al: Molecular signatures of proliferation and quiescence in hematopoietic stem cells. PLoS Biol 2004;2:e301.

16 Akashi K, Traver D, Miyamoto T, Weissman IL: A clonogenic common myeloid progenitor that gives rise to all myeloid lineages. Nature 2000;404: 193–197.

17 Kondo M, Weissman IL, Akashi K: Identification of clonogenic common lymphoid progenitors in mouse bone marrow. Cell 1997;91:661–672.

18 Weissman IL, Shizuru JA: The origins of the identification and isolation of hematopoietic stem cells, and their capability to induce donor-specific transplantation tolerance and treat autoimmune diseases. Blood 2008;112: 3543–3553.

19 Cho RH, Sieburg HB, Muller-Sieburg C: A new mechanism for the aging of hematopoietic stem cells: aging changes the clonal composition of the stem cell compartment but not individual stem cells. Blood 2008;111: 5553–5561.

20 Muller-Sieburg C, Sieburg HB: Stem cell aging: survival of the laziest? Cell Cycle 2008;7:3798–3804.
21 Dykstra B, Kent D, Bowie M, McCaffrey L, Hamilton M, Lyons K, et al: Long-term propagation of distinct hematopoietic differentiation programs in vivo. Cell Stem Cell 2007;1:218–229.
22 Beerman I, Bhattacharya D, Zandi S, Sigvardsson M, Weissman IL, Bryder D, et al: Functionally distinct hematopoietic stem cells modulate hematopoietic lineage potential during aging by a mechanism of clonal expansion. Proc Natl Acad Sci USA 2010;107:5465–5470.
23 Morita Y, Ema H, Nakauchi H: Heterogeneity and hierarchy within the most primitive hematopoietic stem cell compartment. J Exp Med 2010;207:1173–1182.
24 Morrison SJ, Wandycz AM, Akashi K, Globerson A, Weissman IL: The aging of hematopoietic stem cells. Nat Med 1996;2:1011–1016.
25 Sudo K, Ema H, Morita Y, Nakauchi H: Age-associated characteristics of murine hematopoietic stem cells. J Exp Med 2000;192:1273–1280.
26 Rossi DJ, Jamieson CH, Weissman IL: Stems cells and the pathways to aging and cancer. Cell 2008;132:681–696.
27 Wang J, Sun Q, Morita Y, Jiang H, Gross A, Lechel A, et al: A differentiation checkpoint limits hematopoietic stem cell self-renewal in response to DNA damage. Cell 2012;148:1001–1014.
28 Schroeder T: Asymmetric cell division in normal and malignant hematopoietic precursor cells. Cell Stem Cell 2007;1:479–481.
29 Takano H, Ema H, Sudo K, Nakauchi H: Asymmetric division and lineage commitment at the level of hematopoietic stem cells: inference from differentiation in daughter cell and granddaughter cell pairs. J Exp Med 2004;199:295–302.
30 Etienne-Manneville S, Hall A: Rho GTPases in cell biology. Nature 2002;420: 629–635.
31 Florian MC, Dorr K, Niebel A, Daria D, Schrezenmeier H, Rojewski M, et al: Cdc42 activity regulates hematopoietic stem cell aging and rejuvenation. Cell Stem Cell 2012;10:520–530.
32 Brack AS, Conboy MJ, Roy S, Lee M, Kuo CJ, Keller C, et al: Increased Wnt signaling during aging alters muscle stem cell fate and increases fibrosis. Science 2007;317:807–810.
33 Liu H, Fergusson MM, Castilho RM, Liu J, Cao L, Chen J, et al: Augmented Wnt signaling in a mammalian model of accelerated aging. Science 2007;317: 803–806.
34 Atkinson SP, Keith WN: Epigenetic control of cellular senescence in disease: opportunities for therapeutic intervention. Expert Rev Mol Med 2007; 9:1–26.
35 Brena RM, Huang TH, Plass C: Quantitative assessment of DNA methylation: potential applications for disease diagnosis, classification, and prognosis in clinical settings. J Mol Med 2006;84: 365–377.
36 Franke A, DeCamillis M, Zink D, Cheng N, Brock HW, Paro R: Polycomb and polyhomeotic are constituents of a multimeric protein complex in chromatin of *Drosophila melanogaster.* EMBO J 1992;11:2941–2950.
37 Paro R: Propagating memory of transcriptional states. Trends Genet 1995; 11:295–297.
38 Pirrotta V: PcG complexes and chromatin silencing. Curr Opin Genet Dev 1997;7:249–258.
39 Shao Z, Raible F, Mollaaghababa R, Guyon JR, Wu CT, Bender W, et al: Stabilization of chromatin structure by PRC1, a polycomb complex. Cell 1999; 98:37–46.
40 Rada-Iglesias A, Enroth S, Andersson R, Wanders A, Pahlman L, Komorowski J, et al: Histone H3 lysine 27 trimethylation in adult differentiated colon associated to cancer DNA hypermethylation. Epigenetics 2009;4:107–113.
41 Zhang L, Zhong K, Dai Y, Zhou H: Genome-wide analysis of histone H3 lysine 27 trimethylation by ChIP-chip in gastric cancer patients. J Gastroenterol 2009;44:305–312.
42 Gunster MJ, Satijn DP, Hamer KM, den Blaauwen JL, de Bruijn D, Alkema MJ, et al: Identification and characterization of interactions between the vertebrate polycomb-group protein BMI1 and human homologs of polyhomeotic. Mol Cell Biol 1997;17:2326–2335.
43 Hobert O, Jallal B, Ullrich A: Interaction of Vav with ENX-1, a putative transcriptional regulator of homeobox gene expression. Mol Cell Biol 1996;16: 3066–3073.
44 Laible G, Wolf A, Dorn R, Reuter G, Nislow C, Lebersorger A, et al: Mammalian homologues of the Polycomb-group gene Enhancer of zeste mediate gene silencing in *Drosophila* heterochromatin and at *S. cerevisiae* telomeres. EMBO J 1997;16:3219–3232.
45 Satijn DP, Gunster MJ, van der Vlag J, Hamer KM, Schul W, Alkema MJ, et al: RING1 is associated with the polycomb group protein complex and acts as a transcriptional repressor. Mol Cell Biol 1997;17:4105–4113.
46 Schumacher A, Faust C, Magnuson T: Positional cloning of a global regulator of anterior-posterior patterning in mice. Nature 1996;384:648.
47 Sewalt RG, van der Vlag J, Gunster MJ, Hamer KM, den Blaauwen JL, Satijn DP, et al: Characterization of interactions between the mammalian polycomb-group proteins Enx1/EZH2 and EED suggests the existence of different mammalian polycomb-group protein complexes. Mol Cell Biol 1998;18:3586–3595.
48 Tagawa M, Sakamoto T, Shigemoto K, Matsubara H, Tamura Y, Ito T, et al: Expression of novel DNA-binding protein with zinc finger structure in various tumor cells. J Biol Chem 1990;265: 20021–20026.
49 van Lohuizen M, Tijms M, Voncken JW, Schumacher A, Magnuson T, Wientjens E: Interaction of mouse polycomb-group (Pc-G) proteins Enx1 and Enx2 with Eed: indication for separate Pc-G complexes. Mol Cell Biol 1998;18: 3572–3579.
50 van Lohuizen M, Verbeek S, Scheijen B, Wientjens E, van der Gulden H, Berns A: Identification of cooperating oncogenes in E mu-myc transgenic mice by provirus tagging. Cell 1991;65:737–752.
51 Bhattacharyya J, Mihara K, Yasunaga S, Tanaka H, Hoshi M, Takihara Y, et al: BMI-1 expression is enhanced through transcriptional and posttranscriptional regulation during the progression of chronic myeloid leukemia. Ann Hematol 2009;88:333–340.
52 Faubert A, Chagraoui J, Mayotte N, Frechette M, Iscove NN, Humphries RK, et al: Complementary and independent function for Hoxb4 and Bmi1 in HSC activity. Cold Spring Harb Symp Quant Biol 2008;73:555–564.

53 Ohtsubo M, Yasunaga S, Ohno Y, Tsumura M, Okada S, Ishikawa N, et al: Polycomb-group complex 1 acts as an E3 ubiquitin ligase for Geminin to sustain hematopoietic stem cell activity. Proc Natl Acad Sci USA 2008;105:10396–10401.
54 Park IK, Qian D, Kiel M, Becker MW, Pihalja M, Weissman IL, et al: Bmi-1 is required for maintenance of adult self-renewing haematopoietic stem cells. Nature 2003;423:302–305.
55 Kajiume T, Ohno N, Sera Y, Kawahara Y, Yuge L, Kobayashi M: Reciprocal expression of Bmi1 and Mel-18 is associated with functioning of primitive hematopoietic cells. Exp Hematol 2009;37:857–866 e2.
56 Lessard J, Sauvageau G: Bmi-1 determines the proliferative capacity of normal and leukaemic stem cells. Nature 2003;423:255–260.
57 Lukacs RU, Memarzadeh S, Wu H, Witte ON: Bmi-1 is a crucial regulator of prostate stem cell self-renewal and malignant transformation. Cell Stem Cell 2010;7:682–693.
58 van der Lugt NM, Domen J, Linders K, van Roon M, Robanus-Maandag E, te Riele H, et al: Posterior transformation, neurological abnormalities, and severe hematopoietic defects in mice with a targeted deletion of the bmi-1 proto-oncogene. Genes Dev 1994;8:757–769.
59 Molofsky AV, Pardal R, Iwashita T, Park IK, Clarke MF, Morrison SJ: Bmi-1 dependence distinguishes neural stem cell self-renewal from progenitor proliferation. Nature 2003;425:962–967.
60 Pietersen AM, Evers B, Prasad AA, Tanger E, Cornelissen-Steijger P, Jonkers J, et al: Bmi1 regulates stem cells and proliferation and differentiation of committed cells in mammary epithelium. Curr Biol 2008;18:1094–1099.
61 Bruggeman SW, Valk-Lingbeek ME, van der Stoop PP, Jacobs JJ, Kieboom K, Tanger E, et al: Ink4a and Arf differentially affect cell proliferation and neural stem cell self-renewal in Bmi1-deficient mice. Genes Dev 2005;19:1438–1443.
62 Jacobs JJ, Kieboom K, Marino S, DePinho RA, van Lohuizen M: The oncogene and Polycomb-group gene bmi-1 regulates cell proliferation and senescence through the ink4a locus. Nature 1999;397:164–168.
63 Molofsky AV, He S, Bydon M, Morrison SJ, Pardal R: Bmi-1 promotes neural stem cell self-renewal and neural development but not mouse growth and survival by repressing the p16Ink4a and p19Arf senescence pathways. Genes Dev 2005;19:1432–1437.
64 Oguro H, Iwama A, Morita Y, Kamijo T, van Lohuizen M, Nakauchi H: Differential impact of Ink4a and Arf on hematopoietic stem cells and their bone marrow microenvironment in Bmi1-deficient mice. J Exp Med 2006;203:2247–2253.
65 Agherbi H, Gaussmann-Wenger A, Verthuy C, Chasson L, Serrano M, Djabali M: Polycomb mediated epigenetic silencing and replication timing at the INK4a/ARF locus during senescence. PLoS One 2009;4:e5622.
66 Klauke K, Radulovic V, Broekhuis M, Weersing E, Zwart E, Olthof S, et al: Polycomb Cbx family members mediate the balance between haematopoietic stem cell self-renewal and differentiation. Nat Cell Biol 2013;15:353–362.
67 Ringrose L, Paro R: Epigenetic regulation of cellular memory by the Polycomb and Trithorax group proteins. Annu Rev Genet 2004;38:413–443.
68 Hanson RD, Hess JL, Yu BD, Ernst P, van Lohuizen M, Berns A, et al: Mammalian Trithorax and polycomb-group homologues are antagonistic regulators of homeotic development. Proc Natl Acad Sci USA 1999;96:14372–14377.
69 Jacobs JJ, van Lohuizen M: Polycomb repression: from cellular memory to cellular proliferation and cancer. Biochim Biophys Acta 2002;1602:151–161.
70 Lund AH, van Lohuizen M: Epigenetics and cancer. Genes Dev 2004;18:2315–2335.
71 Madan V, Madan B, Brykczynska U, Zilbermann F, Hogeveen K, Dohner K, et al: Impaired function of primitive hematopoietic cells in mice lacking the Mixed-Lineage-Leukemia homolog MLL5. Blood 2009;113:1444–1454.
72 Goll MG, Bestor TH: Eukaryotic cytosine methyltransferases. Annu Rev Biochem 2005;74:481–514.
73 Okano M, Bell DW, Haber DA, Li E: DNA methyltransferases Dnmt3a and Dnmt3b are essential for de novo methylation and mammalian development. Cell 1999;99:247–257.
74 Kaneda M, Okano M, Hata K, Sado T, Tsujimoto N, Li E, et al: Essential role for de novo DNA methyltransferase Dnmt3a in paternal and maternal imprinting. Nature 2004;429:900–903.
75 Challen GA, Sun D, Jeong M, Luo M, Jelinek J, Berg JS, et al: Dnmt3a is essential for hematopoietic stem cell differentiation. Nat Genet 2012;44:23–31.
76 Jeong M, Sun D, Luo M, Huang Y, Challen GA, Rodriguez B, et al: Large conserved domains of low DNA methylation maintained by Dnmt3a. Nat Genet 2014;46:17–23.
77 Pastor WA, Aravind L, Rao A: TETonic shift: biological roles of TET proteins in DNA demethylation and transcription. Nat Rev Mol Cell Biol 2013;14:341–356.
78 Huang S, Law P, Francis K, Palsson BO, Ho AD: Symmetry of initial cell divisions among primitive hematopoietic progenitors is independent of ontogenic age and regulatory molecules. Blood 1999;94:2595–2604.
79 Sanders MA, Valk PJ: The evolving molecular genetic landscape in acute myeloid leukaemia. Curr Opin Hematol 2013;20:79–85.
80 Sun D, Luo M, Jeong M, Rodriguez B, Xia Z, Hannah R, et al: Epigenomic profiling of young and aged HSCs reveals concerted changes during aging that reinforce self-renewal. Cell Stem Cell 2014;14:673–688.
81 Navarro S, Meza NW, Quintana-Bustamante O, Casado JA, Jacome A, McAllister K, et al: Hematopoietic dysfunction in a mouse model for Fanconi anemia group D1. Mol Ther 2006;14:525–535.
82 Nijnik A, Woodbine L, Marchetti C, Dawson S, Lambe T, Liu C, et al: DNA repair is limiting for haematopoietic stem cells during ageing. Nature 2007;447:686–690.
83 Reese JS, Liu L, Gerson SL: Repopulating defect of mismatch repair-deficient hematopoietic stem cells. Blood 2003;102:1626–1633.
84 Rossi DJ, Bryder D, Seita J, Nussenzweig A, Hoeijmakers J, Weissman IL: Deficiencies in DNA damage repair limit the function of haematopoietic stem cells with age. Nature 2007;447:725–729.
85 Janzen V, Forkert R, Fleming HE, Saito Y, Waring MT, Dombkowski DM, et al: Stem-cell ageing modified by the cyclin-dependent kinase inhibitor p16INK4a. Nature 2006;443:421–426.
86 Krishnamurthy J, Ramsey MR, Ligon KL, Torrice C, Koh A, Bonner-Weir S, et al: p16INK4a induces an age-dependent decline in islet regenerative potential. Nature 2006;443:453–457.

87 Molofsky AV, Slutsky SG, Joseph NM, He S, Pardal R, Krishnamurthy J, et al: Increasing p16INK4a expression decreases forebrain progenitors and neurogenesis during ageing. Nature 2006; 443:448–452.
88 Collado M, Blasco MA, Serrano M: Cellular senescence in cancer and aging. Cell 2007;130:223–233.
89 Finkel T, Serrano M, Blasco MA: The common biology of cancer and ageing. Nature 2007;448:767–774.
90 Blasco MA, Lee HW, Hande MP, Samper E, Lansdorp PM, DePinho RA, et al: Telomere shortening and tumor formation by mouse cells lacking telomerase RNA. Cell 1997;91:25–34.
91 Lee HW, Blasco MA, Gottlieb GJ, Horner JW 2nd, Greider CW, DePinho RA: Essential role of mouse telomerase in highly proliferative organs. Nature 1998;392:569–574.
92 Rudolph KL, Chang S, Lee HW, Blasco M, Gottlieb GJ, Greider C, et al: Longevity, stress response, and cancer in aging telomerase-deficient mice. Cell 1999;96:701–712.
93 Choudhury AR, Ju Z, Djojosubroto MW, Schienke A, Lechel A, Schaetzlein S, et al: Cdkn1a deletion improves stem cell function and lifespan of mice with dysfunctional telomeres without accelerating cancer formation. Nat Genet 2007;39:99–105.
94 Ju Z, Jiang H, Jaworski M, Rathinam C, Gompf A, Klein C, et al: Telomere dysfunction induces environmental alterations limiting hematopoietic stem cell function and engraftment. Nat Med 2007;13:742–747.
95 Samper E, Fernandez P, Eguia R, Martin-Rivera L, Bernad A, Blasco MA, et al: Long-term repopulating ability of telomerase-deficient murine hematopoietic stem cells. Blood 2002;99:2767–2775.
96 Begus-Nahrmann Y, Lechel A, Obenauf AC, Nalapareddy K, Peit E, Hoffmann E, et al: Agenetic and genomic analysis identifies a cluster of genes associated with hematopoietic cell turnover. Nat Genet 2009;41:1138–1144.
97 Sperka T, Song Z, Morita Y, Nalapareddy K, Guachalla LM, Lechel A, et al: Puma and p21 represent cooperating checkpoints limiting self-renewal and chromosomal instability of somatic stem cells in response to telomere dysfunction. Nat Cell Biol 2011;14:73–79.
98 Cheng T, Rodrigues N, Dombkowski D, Stier S, Scadden DT: Stem cell repopulation efficiency but not pool size is governed by p27(kip1). Nat Med 2000;6:1235–1240.
99 Cheng T, Rodrigues N, Shen H, Yang Y, Dombkowski D, Sykes M, et al: Hematopoietic stem cell quiescence maintained by p21cip1/waf1. Science 2000;287:1804–1808.
100 van Os R, Kamminga LM, Ausema A, Bystrykh LV, Draijer DP, van Pelt K, et al: A Limited role for p21Cip1/Waf1 in maintaining normal hematopoietic stem cell functioning. Stem Cells 2007;25:836–843.
101 Dunwell TL, Dickinson RE, Stankovic T, Dallol A, Weston V, Austen B, et al: Frequent epigenetic inactivation of the SLIT2 gene in chronic and acute lymphocytic leukemia. Epigenetics 2009;4:265–269.
102 Shibata F, Goto-Koshino Y, Morikawa Y, Komori T, Ito M, Fukuchi Y, et al: Roundabout 4 is expressed on hematopoietic stem cells and potentially involved in the niche-mediated regulation of the side population phenotype. Stem Cells 2009;27:183–190.
103 Gerrits A, Li Y, Tesson BM, Bystrykh LV, Weersing E, Ausema A, et al: Expression quantitative trait loci are highly sensitive to cellular differentiation state. PLoS Genet 2009; 5:e1000692.
104 Bee T, Ashley E, Bickley S, Jarrat A, Li P-S, Sloane-Stanley J, et al: The mouse Runx1 23 hematopoietic stem cell enhancer confers hematopoietic specificity to both Runx1 promoters. Blood 2009;113:5121–5124.
105 Mohn F, Schübeler D: Genetics and epigenetics: stability and plasticity during cellular differentiation. Trends Genet 2009;25:129–136.
106 Yilmaz OH, Valdez R, Theisen BK, Guo W, Ferguson DO, Wu H, et al: Pten dependence distinguishes haematopoietic stem cells from leukaemia-initiating cells. Nature 2006;441: 475–482.
107 Zhang J, Grindley JC, Yin T, Jayasinghe S, He XC, Ross JT, et al: PTEN maintains haematopoietic stem cells and acts in lineage choice and leukaemia prevention. Nature 2006;441: 518–522.
108 Wilson A, Murphy MJ, Oskarsson T, Kaloulis K, Bettess MD, Oser GM, et al: c-Myc controls the balance between hematopoietic stem cell self-renewal and differentiation. Genes Dev 2004;18:2747–2763.
109 Yuan Y, Shen H, Franklin DS, Scadden DT, Cheng T: In vivo self-renewing divisions of haematopoietic stem cells are increased in the absence of the early G1-phase inhibitor, p18INK4C. Nat Cell Biol 2004;6:436–442.
110 Beerman I, Bock C, Garrison BS, Smith ZD, Gu H, Meissner A, et al: Proliferation-dependent alterations of the DNA methylation landscape underlie hematopoietic stem cell aging. Cell Stem Cell 2013;12:413–425.
111 Yu H, Yuan Y, Shen H, Cheng T: Hematopoietic stem cell exhaustion impacted by p18 INK4C and p21 Cip1/ Waf1 in opposite manners. Blood 2006;107:1200–1206.
112 Lacorazza HD, Nimer SD: The emerging role of the myeloid Elf-1 like transcription factor in hematopoiesis. Blood Cells Mol Dis 2003;31: 342–350.
113 Lacorazza HD, Yamada T, Liu Y, Miyata Y, Sivina M, Nunes J, et al: The transcription factor MEF/ELF4 regulates the quiescence of primitive hematopoietic cells. Cancer Cell 2006;9: 175–187.
114 Min IM, Pietramaggiori G, Kim FS, Passegue E, Stevenson KE, Wagers AJ: The transcription factor EGR1 controls both the proliferation and localization of hematopoietic stem cells. Cell Stem Cell 2008;2:380–391.
115 Wilson A, Laurenti E, Oser G, van der Wath RC, Blanco-Bose W, Jaworski M, et al: Hematopoietic stem cells reversibly switch from dormancy to self-renewal during homeostasis and repair. Cell 2008;135:1118–1129.
116 Rodgers JT, King KY, Brett JO, Cromie MJ, Charville GW, Maguire KK, et al: mTORC1 controls the adaptive transition of quiescent stem cells from G0 to G(Alert). Nature 2014;509: 393–396.
117 Yamazaki S, Iwama A, Takayanagi S, Morita Y, Eto K, Ema H, et al: Cytokine signals modulated via lipid rafts mimic niche signals and induce hibernation in hematopoietic stem cells. EMBO J 2006;25:3515–3523.

118 Yamazaki S, Iwama A, Takayanagi S, Eto K, Ema H, Nakauchi H: TGF-beta as a candidate bone marrow niche signal to induce hematopoietic stem cell hibernation. Blood 2009;113:1250–1256.
119 Borden EC, Sen GC, Uze G, Silverman RH, Ransohoff RM, Foster GR, et al: Interferons at age 50: past, current and future impact on biomedicine. Nat Rev Drug Discov 2007;6:975–990.
120 Darnell JE Jr, Kerr IM, Stark GR: Jak-STAT pathways and transcriptional activation in response to IFNs and other extracellular signaling proteins. Science 1994;264:1415–1421.
121 Kujawski LA, Talpaz M: The role of interferon-alpha in the treatment of chronic myeloid leukemia. Cytokine Growth Factor Rev 2007;18:459–471.
122 Stark GR, Kerr IM, Williams BR, Silverman RH, Schreiber RD: How cells respond to interferons. Annu Rev Biochem 1998;67:227–264.
123 Essers MA, Offner S, Blanco-Bose WE, Waibler Z, Kalinke U, Duchosal MA, et al: IFNalpha activates dormant haematopoietic stem cells in vivo. Nature 2009;458:904–908.

Yohei Morita
Leibniz Institute for Age Research, Fritz Lipmann Institute e.V. (FLI)
Beutenbergstrasse 11
DE–07745 Jena (Germany)
E-Mail ymorita@fli-leibniz.de

Rudolph KL (ed): Adult Stem Cells in Aging, Diseases and Cancer.
Else Kröner-Fresenius Symp. Basel, Karger, 2015, vol 5, pp 60–65 (DOI: 10.1159/000366569)

The Microenvironment, Aging and Disease

Hartmut Geiger

Institute of Molecular Medicine, Aging and Stem Cells, University of Ulm, Ulm, Germany

Abstract

Stem cell function declines with age. It has been hypothesized that aging of stem cells is the underlying cause of impaired tissue homeostasis as well as cancer in aged individuals. In 1978, Schofield proposed the 'niche' hypothesis to describe the physiologically limited microenvironment that supports stem cells. The role of niche aging for stem cell aging is still not understood in detail. A central question in stem cell aging research is consequently the identification of the contribution of stem cell-extrinsic factors (including the niche and/or systemic factors) to stem cell aging, especially in the light of aging-associated diseases.

The Influence of the Niche and the Environment on Stem Cells in Disease and Aging

Stem cell function declines with age in both humans and mice [1–7]. It has been hypothesized that aging of stem cells is the underlying cause of impaired tissue homeostasis as well as cancer in aged individuals which might ultimately limit lifespan [1, 2, 8–11].

In 1978, Schofield proposed the 'niche' hypothesis to describe the physiologically limited microenvironment that supports stem cells. This hypothesis has been confirmed in multiple stem cell systems from distinct organisms over the last decade, and is currently an active field of research in stem cell biology [12, 13]. In recent years, the aspect of aging of the niche on altered phenotypes associated with aged stem cells is more and more appreciated [14]. One central question in stem cell and stem cell aging research is consequently the identification of the contribution of stem cell-intrinsic versus -extrinsic factors (including the niche and/or systemic factors) to their biology, which is critically important with respect to therapeutic interventions. The following chapter highlights novel insights from the 5th Else Kröner-Fresenius Symposium on Adult Stem Cells in Aging, Diseases and Cancer, focusing primarily on extrinsic and niche factors that influence stem cells and aging, from Drs. Trumpp, Jasper, and Geiger.

Modeling Myelodysplastic Syndromes: The Role of the Niche

Aging is a major risk factor for leukemia, with an exponential increase in the incidence with age. The biological mechanisms of this age-associated increased incidence are not known in detail, but it is widely accepted that the combination of intrinsic and extrinsic mechanisms with respect to leukemia-initiating cells drives this process. While likely hematopoietic cell intrinsic mechanisms that might contribute to the increase in leukemia with age have been described [3, 12], whether aging of the BM microenvironment also influences pre-leukemic states or leukemia has not been studied in great detail so far [4]. The marrow microenvironment is able to influence differentiation [5], drug resistance [6] and the proliferation of leukemic cells in diverse mouse leukemia models [7] while it undergoes multiple changes upon aging as demonstrated by decreased bone remodeling [9], enhanced adipogenesis and changes in ECM components [10, 11]. This might be a consequence of the aging of mesenchymal stem cells (MSCs), the stem cells that at least in part are the underlying resource for cells forming the stem cell niche. Myelodysplastic syndrome (MDS) constitutes a complex set of hematopoietic stem cell diseases that involves defective differentiation into either one, some or all of the hematopoietic lineages. MDS is caused by the accumulation of mutations in hematopoietic stem/progenitor cells (HSPCs) that result in the accumulation of defective HSPCs in the bone marrow.

MDS is one of the most aging-associated diseases of the hematopoietic system, with an exponential increase in incidence up from the age of 50, with a median age of about 70. It has also been demonstrated by the laboratory of David Scadden in a murine leukemia resembling some aspects of MDS that changes in the niche/microenvironment can cause MDS [15]. Thus, the question arises to which extent stroma and especially aged stroma in aged MDS patients might contribute to MDS. The laboratory of Dr. Trumpp thus investigates the contribution of niche/stroma components on the MDS progression in a murine xenotransplant model. Dr. Trumpp asked in his presentation whether it might be possible that an MDS clone forms a functional unit together with its niche. Using a novel kind of mouse strain called NOD/SCID IL2-RG (NSG)-S initially reported on by the laboratory of James Mulloy from Cincinnati [16], in which human hematopoietic cytokines (SCF, GM-CSF and IL-3) are provided as transgenes to allow for a more humanized niche compared to standard NSG animals. Relying on a set of early-stage MDS samples provided from the University Hospital in Mannheim, Germany, he demonstrated that cotransplantation of patient-derived MSCs (defined as $CD45^{-}HLA\text{-}DR^{-}CD105^{+}$ $CD73^{+}CD90^{+}CD44^{+}CD146^{+}$ cells) support engraftment of the MDS disease. The MDS clones usually grow better in the NSG-S recipients compared to NSG recipients. Moreover, MDS typical features such as the appearance of megaloblastic and vacuolized proerythroblasts were observed only in the former. These data provide evidence that MDS-derived MSCs serve as a niche, strongly influencing disease progression. Initial gene expression experiments actually confirmed that diseased MSCs are distinct from normal age-matched MSCs. Further experiments presented also suggested that patient-derived MSCs provided at the same time as MDS hematopoietic cells promote higher engraftment than age-matched healthy MSCs in the NSG-S MDS model. Finally, data were presented suggesting that MDS cells reprogram MSCs in order to express cytokines and other factors that promote MDS cell growth and probably inhibit normal hematopoiesis. In summary, the data presented indicated that MDS stem cells require autologous niche cells (MSCs) for disease initiation and propagation [17–20].

Stem Cell Proliferation and Tissue Homeostasis in the Aging *Drosophila* Intestine

Tissue homeostasis is driven by the balance between regeneration and dysplasia and/or cancer. In aging, ROS, tissue damage and inflammation are thought to influence this balance negatively. The laboratory of Dr. Jasper studies the intestinal epithelium of flies *(Drosophila melanogaster).* They are focused on the role of intestinal stem cells (ISCs) in the aging of the fly gut. The *Drosophila* intestine is a productive model to characterize age-related changes in host-commensal interactions, innate immune signaling, and regenerative capacity [21]. The intestinal epithelium is constantly being renewed in a process that mechanistically and morphologically resembles cell turnover in the intestinal epithelium of mammals [21, 22]. ISCs are the only dividing cells in the intestinal epithelium and can give rise, when activated and not quiescent, to at least two differentiated intestinal cell types: enteroendocrine cells and enterocytes (ECs) [21, 22]. This lineage is deregulated in the aging intestine where ISC proliferation strongly increases, and polyploid, misdifferentiated cells accumulate in the epithelium [23, 24]. This dysplastic phenotype disrupts epithelial function, resulting in age-related metabolic decline in the whole fly, as well as loss of the intestinal barrier function [23, 25, 26]. Intestinal dysplasia influences *Drosophila* lifespan as limiting the rate of ISC proliferation in the aging intestine is sufficient to extend lifespan [23, 25]. The signaling mechanisms that contribute to that overproliferation are the JAK/STAT, p38 and JNK signaling pathways [27]. Some of these pathways are driven by the cytokine interleukin-6, which is usually associated with inflammation. The proteins Nrf2 and FOXO actually limit the aging-associated overproliferation. Overexpression of target genes of NRF2 and FOXO leads to lifespan extension [28].

The Jasper lab is also interested in the role of innate immune signaling in the gut. Recent work in the Jasper lab, which has now been published in *Cell* [29], shows that the age-related loss of innate immune homeostasis is the underlying cause of intestinal dysplasia. The gut of old flies presents with an expansion of commensal bacteria, which correlates with elevated ISC proliferation. Is there a causal relationship? Yes, as an axenic condition (sterile flies without commensal bacteria) prevents the age-related increase in ISC proliferation. Transcriptome profiles of axenic and control intestinal tissue revealed pathways that are influenced by the commensal bacteria. These include oxidative response genes, which is consistent with an increase in ROS production by intestinal dual oxidase as a response to the age-related commensal expansion.

Interestingly, FOXO was chronologically activated in ECs of both axenic and conventionally aging flies, and this chronic activation of FOXO results in high Rel/NF-κB activity in aging ECs. Accordingly, knockdown of FOXO in ECs prevents the induction of inflammatory markers upon aging. The Jasper lab has explored the mechanism of the interaction between FOXO and the Rel/NF-κB signaling pathway and found that FOXO regulates the expression of peptidoglycan recognition proteins of the SC class (PGRP-SCs), which are negative regulators of IMD/Rel signaling. Overexpression of PGRP-SC2 in ECs decimated the number of commensal bacteria in old flies and restored ISC quiescence, while also significantly extending lifespan of flies.

In summary, these novel and exciting data support that in young animals, PGRP-SC2 maintains immune homeostasis by limiting IMD/Rel immune deficiency pathway activity. Age-associated chronic activation of Foxo in ECs (for which the mechanism is still ill-defined) results in repression of PGRP-SC2 expression and deregulation of IMD/Rel signaling. The associated immunosenescence causes commensal dysbiosis, which chronically activates ROS production, triggering ISC overproliferation and dysplasia.

Polarity and Wnt5a Signaling Are Central to Stem Cell Aging

Aging results in the inability of tissues to maintain homeostasis, and it is believed that somatic stem cell aging is one underlying cause of tissue attrition with age or age-related diseases. Aging of hematopoietic stem cells (HSCs) is associated with impaired hematopoiesis in the elderly. Despite a large amount of data describing the decline of HSC function upon aging, the molecular mechanisms of this process remain still largely unknown, which precludes rational approaches to attenuate stem cell aging.

Dr. Geiger's laboratory previously published that primitive hematopoietic cells (early hematopoietic progenitor cells, LIN$^-$, Sca-1$^+$, C-Kit$^+$: eHPCs) in aged mice show reduced adhesion to stromal cells correlating with an increased activity of the small RhoGTPase Cdc42 compared to eHPCs from young mice, but no change in the activity of Rac1 and Rac2 [30]. In collaboration with Dr. Gunzer in Magedburg, Germany, time-lapse multi-photon intravital microscopy was used to determine the dynamics of young and aged hematopoietic cells inside the diaphysis of a long bone inside the marrow cavity. eHPCs in young mice, although remaining fixed on the spot, presented with constantly ongoing cell protrusion movements indicating synapse-like interactions with stromal cells. In contrast, eHPCs in aged mice displayed a higher cell protrusion activity and were localized more distantly from the endosteum compared to young eHPCs, while showing reduced adhesion to stromal cells and presenting with a reduced number of polarized cells upon adhesion. These intracellular changes are in part regulated by outside-in signaling of receptors sensing the cellular environment that then results in inside-out reactions of the cell and either adhesion or detachment. This outside-in information is transmitted through members of the family of small Rho GTPases and other signaling pathways [31–33]. Small Rho GTPases, including Rac1, Rac2, Cdc42 and RhoA are primarily involved in the regulation of adhesion signaling in response to outside signals in various mammalian cell types including hematopoietic cells, and the individual contribution of these small Rho GTPases to regulation is cell type specific as well as adhesion substrate specific [34–39].

Thus, critical changes in HSC function with aging might be a consequence of the increased activity of Cdc42 in aged HSCs, which goes along with novel findings of Dr. Geiger showing that HSCs from young animals with genetically elevated activity of Cdc42 functionally present as aged HSCs. To further test their hypothesis, Dr. Geiger's lab isolated young and aged long-term repopulating HSCs from bone marrow of C57Bl/6 mice and analyzed expression and localization of Cdc42 together with actin, tubulin and several polarity proteins and adhesion receptors. Single-cell immunofluorescence staining showed that Cdc42 and tubulin localized in a highly polarized way in young HSCs. Coimmunostaining of Cdc42 and tubulin revealed that both proteins accumulate at the same pole of the cell in young HSCs, while only a small percentage of aged HSCs showed a polar distribution of Cdc42 and tubulin. Pharmacological inhibition of Cdc42 activity significantly decreased the percentage of young HSCs polarized for Cdc42 and tubulin and reverted the apolar distribution of Cdc42 and tubulin of aged HSCs in a dose-dependent manner, indicating that Cdc42 activity controls the polar localization of Cdc42 in HSCs and that polarization of both tubulin and Cdc42 itself depends on Cdc42 activity.

Aged muscle stem cells, in which canonical Wnt signaling is elevated, can be activated to differentiate and regenerate muscles in aged mice as efficiently as young muscle stem cells either by forced activation of Notch or by Wnt inhibitor factors in serum from young animals supplied by parabiosis [3, 40–42]. Whether there is a similar critical role of Wnt signaling in aging of HSCs remains largely unexplored. Wnt signaling was previously associated with Cdc42 signaling. The

Geiger laboratory next determined expression levels of several Wnt mRNAs in young and aged (24- to 26-month-old C57/BL6 mice) HSCs by real-time RT-PCR. Surprisingly, aged HSCs (LIN$^-$, Sca-1$^+$, c-Kit$^+$, CD34$^-$, flk2$^-$, from 24- to 26-month-old C57/BL6 mice) presented with high levels of expression of Wnt5a and Wnt4, primarily associated with noncanonical Wnt signaling, while absent in young HSCs (from 10- to 12-week-old C57/BL6 mice). In contrast, expression of Wnt1, 3a, 5b and 10b, which are associated with canonical Wnt signaling, was not altered upon HSC aging. Axin2, a direct downstream target of canonical Wnt signaling, was markedly decreased in aged HSCs alongside with reduced expression and reduced nuclear localization of β-catenin, implying an inhibitory role for noncanonical Wnt signaling on canonical Wnt signaling upon HSC aging. We recently reported that the majority of young HSCs are polar for a large number of planar cell polarity proteins and tubulin as well as for acetylated HistoneH4 at Lysine16 within the nucleus (AcH4K16), while the majority of aged HSCs are apolar.

Implying a causal role for Wnt5a in aging of HSCs, treatment of young HSCs induced (1) aging-associated stem cell apolarity, (2) reduction of regenerative capacity and (3) an aging-like myeloid-lymphoid differentiation skewing via activating the small RhoGTPase Cdc42. Conversely, Wnt5a haploinsufficiency (Wnt5a$^{+/-}$ 24-month-old C57/Bl6 mice; Wnt5a$^{-/-}$ mice die perinatally) attenuated HSC aging (restoration of HSC polarity, increased level of B cell and reduction of myeloid cells in peripheral blood, increased red blood cell count and hemoglobin levels), while stem cell intrinsic reduction of Wnt5a expression by a knockdown approach in already aged HSCs results in functionally rejuvenated aged HSCs, as measured in functional transplantation experiments.

In summary, these data demonstrate an unexpected shift from canonical to noncanonical Wnt signaling due to, at least in part, stem cell intrinsically elevated Wnt5a expression in aged HSCs. Mechanisms inducing elevated expression of Wnt5a in aged HSCs possibly involve epigenetic mechanisms as the level of Wnt5a expression in other cells is malleable by epigenetic regulation.

References

1 Van Zant G, Liang Y: The role of stem cells in aging. Exp Hematol 2003;31: 659–672.
2 Geiger H, Van Zant G: The aging of lympho-hematopoietic stem cells. Nat Immunol 2002;3:329–333.
3 Conboy IM, Conboy MJ, Wagers AJ, Girma ER, Weissman IL, Rando TA: Rejuvenation of aged progenitor cells by exposure to a young systemic environment. Nature 2005;433:760–764.
4 Kirkwood TB: Intrinsic ageing of gut epithelial stem cells. Mech Ageing Dev 2004;125:911–915.
5 Potten CS, Martin K, Kirkwood TB: Ageing of murine small intestinal stem cells. Novartis Found Symp 2001;235: 66–79; discussion 79–84, 101–104.
6 Kim M, Moon HB, Spangrude GJ: Major age-related changes of mouse hematopoietic stem/progenitor cells. Ann NY Acad Sci 2003;996:195–208.
7 Sudo K, Ema H, Morita Y, Nakauchi H: Age-associated characteristics of murine hematopoietic stem cells. J Exp Med 2000;192:1273–1280.
8 Geiger H, True JM, De Haan G, Van Zant G: Longevity and stem cells: a genetic connection. Sci World J 2001;1(suppl 3):77.
9 Sharpless NE, DePinho RA: Telomeres, stem cells, senescence, and cancer. J Clin Invest 2004;113:160–168.
10 Torella D, Rota M, Nurzynska D, Musso E, Monsen A, Shiraishi I, Zias E, Walsh K, Rosenzweig A, Sussman MA, Urbanek K, Nadal-Ginard B, Kajstura J, Anversa P, Leri A: Cardiac stem cell and myocyte aging, heart failure, and insulin-like growth factor-1 overexpression. Circ Res 2004;94:514–524.
11 Geiger H, de Haan G, Florian MC: The ageing haematopoietic stem cell compartment. Nat Rev Immunol 2013;13: 376–389.
12 Li L, Xie T: Stem cell niche: structure and function. Annu Rev Cell Dev Biol 2005;21:605–631.
13 Papayannopoulou T, Scadden DT: Stem-cell ecology and stem cells in motion. Blood 2008;111:3923–3930.
14 Ju Z, Jiang H, Jaworski M, Rathinam C, Gompf A, Klein C, Trumpp A, Rudolph KL: Telomere dysfunction induces environmental alterations limiting hematopoietic stem cell function and engraftment. Nat Med 2007;13:742–747.
15 Raaijmakers MH, Mukherjee S, Guo S, Zhang S, Kobayashi T, Schoonmaker JA, Ebert BL, Al-Shahrour F, Hasserjian RP, Scadden EO, Aung Z, Matza M, Merkenschlager M, Lin C, Rommens JM, Scadden DT: Bone progenitor dysfunction induces myelodysplasia and secondary leukaemia. Nature 2010;464: 852–857.

16 Wunderlich M, Chou FS, Link KA, Mizukawa B, Perry RL, Carroll M, Mulloy JC: AML xenograft efficiency is significantly improved in NOD/SCID-IL2RG mice constitutively expressing human SCF, GM-CSF and IL-3. Leukemia 2010;24:1785–1788.
17 Medyouf H, Mossner M, Jann JC, Nolte F, Raffel S, Herrmann C, Lier A, Eisen C, Nowak V, Zens B, Mudder K, Klein C, Oblander J, Fey S, Vogler J, Fabarius A, Riedl E, Roehl H, Kohlmann A, Staller M, Haferlach C, Muller N, John T, Platzbecker U, Metzgeroth G, Hofmann WK, Trumpp A, Nowak D: Myelodysplastic cells in patients reprogram mesenchymal stromal cells to establish a transplantable stem cell niche disease unit. Cell Stem Cell 2014;14:824–837.
18 Katayama Y, Battista M, Kao WM, Hidalgo A, Peired AJ, Thomas SA, Frenette PS: Signals from the sympathetic nervous system regulate hematopoietic stem cell egress from bone marrow. Cell 2006;124:407–421.
19 Frenette PS, Pinho S, Lucas D, Scheiermann C: Mesenchymal stem cell: keystone of the hematopoietic stem cell niche and a stepping-stone for regenerative medicine. Annu Rev Immunol 2013;31:285–316.
20 Yamazaki S, Ema H, Karlsson G, Yamaguchi T, Miyoshi H, Shioda S, Taketo MM, Karlsson S, Iwama A, Nakauchi H: Nonmyelinating Schwann cells maintain hematopoietic stem cell hibernation in the bone marrow niche. Cell 2011;147:1146–1158.
21 Buchon N, Broderick NA, Lemaitre B: Gut homeostasis in a microbial world: insights from *Drosophila melanogaster.* Nat Rev Microbiol 2013;11:615–626.
22 Biteau B, Hochmuth CE, Jasper H: Maintaining tissue homeostasis: dynamic control of somatic stem cell activity. Cell Stem Cell 2011;9:402–411.
23 Biteau B, Karpac J, Supoyo S, Degennaro M, Lehmann R, Jasper H: Lifespan extension by preserving proliferative homeostasis in *Drosophila*. PLoS Genet 2010;6:e1001159.
24 Biteau B, Hochmuth CE, Jasper H: JNK activity in somatic stem cells causes loss of tissue homeostasis in the aging *Drosophila* gut. Cell Stem Cell 2008;3:442–455.
25 Rera M, Bahadorani S, Cho J, Koehler CL, Ulgherait M, Hur JH, Ansari WS, Lo T Jr, Jones DL, Walker DW: Modulation of longevity and tissue homeostasis by the *Drosophila* PGC-1 homolog. Cell Metab 2011;14:623–634.
26 Rera M, Clark RI, Walker DW: Intestinal barrier dysfunction links metabolic and inflammatory markers of aging to death in *Drosophila*. Proc Natl Acad Sci USA 2012;109:21528–21533.
27 Biteau B, Karpac J, Hwangbo D, Jasper H: Regulation of *Drosophila* lifespan by JNK signaling. Exp Gerontol 2011;46:349–354.
28 Kapuria S, Karpac J, Biteau B, Hwangbo D, Jasper H: Notch-mediated suppression of TSC2 expression regulates cell differentiation in the *Drosophila* intestinal stem cell lineage. PLoS Genet 2012;8:e1003045.
29 Guo L, Karpac J, Tran SL, Jasper H: PGRP-SC2 promotes gut immune homeostasis to limit commensal dysbiosis and extend lifespan. Cell 2014;156:109–122.
30 Xing Z, Ryan MA, Daria D, Nattamai KJ, Van Zant G, Wang L, Zheng Y, Geiger H: Increased hematopoietic stem cell mobilization in aged mice. Blood 2006;108:2190–2197.
31 Hall A: G proteins and small GTPases: distant relatives keep in touch. Science 1998;280:2074–2075.
32 Hall A: Rho GTPases and the actin cytoskeleton. Science 1998;279:509–514.
33 Defilippi P, Valles AM: The winding road from adhesive receptors to the nucleus. Conference on molecular biology of cellular interactions: adhesion receptor signalling and regulation of gene expression. EMBO Rep 2002;3:312–316.
34 Yang FC, Atkinson SJ, Gu Y, Borneo JB, Roberts AW, Zheng Y, Pennington J, Williams DA: Rac and Cdc42 GTPases control hematopoietic stem cell shape, adhesion, migration, and mobilization. Proc Natl Acad Sci USA 2001;98:5614–5618.
35 Gu Y, Filippi MD, Cancelas JA, Siefring JE, Williams EP, Jasti AC, Harris CE, Lee AW, Prabhakar R, Atkinson SJ, Kwiatkowski DJ, Williams DA: Hematopoietic cell regulation by Rac1 and Rac2 guanosine triphosphatases. Science 2003;302:445–449.
36 Jansen M, Yang FC, Cancelas JA, Bailey JR, Williams DA: Rac2-deficient hematopoietic stem cells show defective interaction with the hematopoietic microenvironment and long-term engraftment failure. Stem Cells 2005;23:335–346.
37 Cancelas JA, Lee AW, Prabhakar R, Stringer KF, Zheng Y, Williams DA: Rac GTPases differentially integrate signals regulating hematopoietic stem cell localization. Nat Med 2005;11:886–891.
38 Cancelas JA, Jansen M, Williams DA: The role of chemokine activation of Rac GTPases in hematopoietic stem cell marrow homing, retention, and peripheral mobilization. Exp Hematol 2006;34:976–985.
39 Ghiaur G, Lee A, Bailey J, Cancelas J, Zheng Y, Williams DA: Inhibition of RhoA GTPase activity enhances hematopoietic stem and progenitor cell proliferation and engraftment in vivo. Blood 2006;108:2087–2094.
40 Conboy IM, Conboy MJ, Smythe GM, Rando TA: Notch-mediated restoration of regenerative potential to aged muscle. Science 2003;302:1575–1577.
41 Brack AS, Conboy MJ, Roy S, Lee M, Kuo CJ, Keller C, Rando TA: Increased Wnt signaling during aging alters muscle stem cell fate and increases fibrosis. Science 2007;317:807–810.
42 Liu H, Fergusson MM, Castilho RM, Liu J, Cao L, Chen J, Malide D, Rovira II, Schimel D, Kuo CJ, Gutkind JS, Hwang PM, Finkel T: Augmented Wnt signaling in a mammalian model of accelerated aging. Science 2007;317:803–806.

Hartmut Geiger, PhD
Institut für Molekulare Medizin
Albert-Einstein Allee
DE–89081 Ulm (Germany)
E-Mail hartmut.geiger@uni-ulm.de

Rudolph KL (ed): Adult Stem Cells in Aging, Diseases and Cancer.
Else Kröner-Fresenius Symp. Basel, Karger, 2015, vol 5, pp 66–68 (DOI: 10.1159/000366571)

Sestrins in Aging and Metabolism

Duozhuang Tang • Si Tao

Leibniz Institute for Age Research, Fritz Lipmann Institute, Jena, Germany

Abstract

During the 5th Else Kröner-Fresenius Symposium, new data were presented on the role of sestrins (Sesns) in regulating metabolism and therefore the development of aging-related pathologies. Sesns are a highly conserved gene family among eukaryotes. There are three Sesns (Sesn1, Sesn2 and Sesn3) in mammals but one ortholog in *Drosophila*. Sesns are regulated by many stress insults including DNA damage, oxidative stress, hypoxia, growth factor deprivation, hyperactive TOR and RAS signaling via p53 and FOXOs, and result in enhanced autophagy, reactive oxygen species reduction, reduced protein synthesis, reduced anabolism, and reduced cell growth. These Sesn-dependent effects are conserved in *Drosophila* and in mammalian cells. Sesns are inhibitors of TOR signaling. Nutritional abundance leads to TORC1 activation, thus enhancing biosynthesis of protein and lipid through p70S6K and SREBP, and inhibiting autophagy through ULK1. Continuous activation of TOR is associated with various pathologies, including obesity-associated diseases, declined cardiac function, and cancer.

Metabolism is influenced by multiple factors, including age, food intake, and conditions such as diabetes and obesity [1]. Energy metabolism and energy-sensing pathways mainly involve AMPK (AMP-activated protein kinase), FOXO (insulin-forkhead box O factors), TOR (target of rapamycin), SIRT (sirtuins), and insulin signaling [2–4]. TOR plays an important role in regulating cell growth and maintaining cellular and organism homeostasis [5–7]. Different status of metabolism of the cell and its niche could lead to activation or inactivation of these signaling pathways, and consequently influence cell fate regulation. Emerging evidence shows that stem cells have distinct metabolism compared to differentiated cells, and the unique metabolism property of stem cells is important for their maintenance. Energy metabolism plays an important role in tissue homeostasis and therefore is of great interest for studies in aging as well as tumorigenesis. In this chapter, we will review molecular regulators of metabolic pathways that were discussed at the 5th Else Kröner-Fresenius Symposium on Stem Cells in Aging, Diseases and Cancer.

Drosophila Sestrin as a Feedback Inhibitor of TOR Protects *Drosophila* from Accumulation of Aging-Related Pathologies

Both RNA and protein of *Drosophila Sestrin* (dSesn) are increased upon persistent TOR activation through accumulation of reactive oxygen species (ROS) which cause activation of c-Jun and

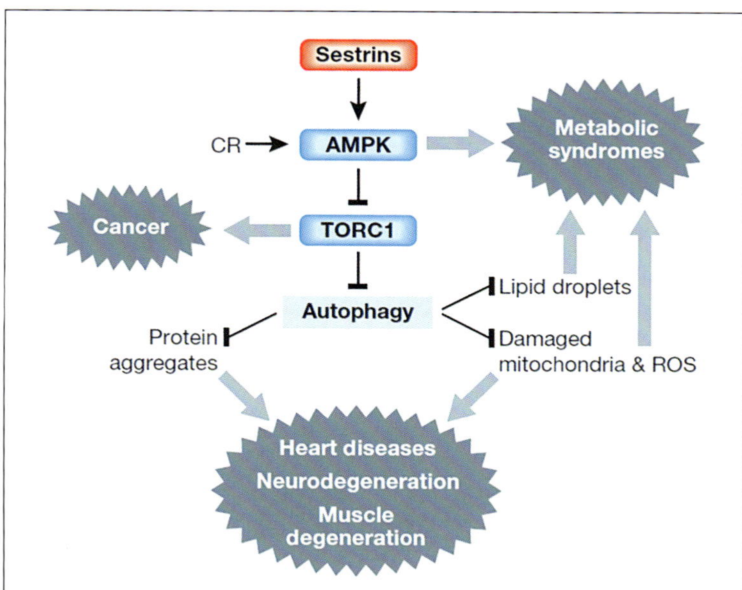

Fig. 1. Potential role of Sesns in aging and age-associated diseases. Both Sesns and CR (calorie restriction) can activate AMPK, inhibit TORC1 and induce autophagy, which are considered to suppress aging and age-associated disease processes [12].

FoxO. Using dSesn-null flies as a model, three major functions of dSesn were identified: (1) suppression of lipid accumulation; (2) prevention of cardiac malfunction, and (3) protection of muscle from age-related degeneration. dSesn-null flies show increased accumulation of lipids, which can be reduced by ectopic expression of dSesn. dSesn also controls carbohydrate metabolism [8]. Relative concentration of hemolymph trehalose (a major blood sugar of insects) significantly increases in dSesn-null flies. Age-related decline in cardiac function is associated with TOR hyperactivity [9–11]. dSesn-null mutants show arrhythmia and decreased heart rate, and structural disorganization of myofibrils, which is rescued by TOR inhibition, suggesting that dSesn provides cardioprotection by inhibiting TOR activity. dSesn-null flies also exhibited degeneration of thoracic muscles and mitochondrial abnormalities leading to excessive generation of ROS and muscle cell death. These findings suggest a role of dSesn in protecting thoracic muscles against age-dependent progressive degeneration. Interestingly, dSesn deficiency results in oxidative stress and accumulation of protein aggregates which are hallmarks of defective autophagy. Inhibition of autophagy by silencing its essential component ATG1 phenocopied dSesn loss of function. dSesn loss causes upregulation of TOR, which inhibits physiological autophagy resulting in a decrease in elimination of dysfunctional mitochondria and therefore contributes to muscle degeneration. In summary, this study showed that dSesn serves as a negative feedback regulator of TOR. dSesn loss of function results in hyperactive TOR which contributes to several biochemical and cell biological defects associated with aging [8].

Sesn2 and Sesn3 Maintain Metabolic Homeostasis in Mammalians

To understand the potential role of Sesns in metabolism and age-related diseases in mammals, Budanov and colleagues studied obesity-associated metabolic pathologies in Sesn-deficient mice lacking Sesn3 and/or Sesn2 ($Sesn2^{-/-}$ and $Sesn3^{-/-}$ mice). Expression of Sesn2 (mRNA and protein) is elevated in liver of mice kept on high-fat diet (HFD) which is a model of obesity and type 2 diabetes, while the expression of Sesn1 and Sesn3 is not changed [12]. $Sesn2^{-/-}$ mice do not show abnormalities in glucose homeostasis and insulin responsiveness when maintained on normal chow; however, they show a significantly higher degree of glucose intolerance and insulin resistance in HFD mice [12]. Inactivation of Sesn2 also exacerbates diabetic phenotype of $Lep^{ob/ob}$ mice, a genetic model of obesity. Sesns regulate AKT, an important effector of insulin signaling in the liver, in an AMPK-, TSC2- and TORC2-dependent manner. Alterations in glucose homeostasis in $Sesn2^{-/-}$ mice on HFD were accompanied by suppression of AMPK activity, which compromised PI3K-AKT signaling (attenuating insulin-induced AKT activity) as a result of enhanced mTORC1 activation [12]. In support of this interpretation, AMPK reactivation restored insulin sensitivity in Sesn2 KO mice. Sesns also play an important role in lipid metabolism. Sesn2 deficiency stimulates lipid accumulation and attenuates fatty acid β-oxidation, lipolysis, and autophagy in the liver of HFD mice and $Lep^{ob/ob}$ mice. Concomitant loss of Sesn2 and Sesn3 resulted in mTORC1 activation and spontaneous insulin resistance in the liver even in the absence of nutritional overload and obesity. These results demonstrate an important role of mammalian Sesns in the control of lipid and glucose metabolism [12]. Interestingly, new data from the Budanov group indicate that Sesns can promote or suppress carcinogenesis dependent on the context.

Taken together, the above studies suggest a potential role of Sesns in modulating aging and age-related diseases (fig. 1). Sesns may prevent aging and age-associated pathologies by activating AMPK, inhibiting TORC1 and inducing autophagy [13].

References

1 Rafalski VA, Mancini E, Brunet A: Energy metabolism and energy-sensing pathways in mammalian embryonic and adult stem cell fate. J Cell Sci 2012;125:5597–5608.

2 Greer EL, Brunet A: FOXO transcription factors at the interface between longevity and tumor suppression. Oncogene 2005;24:7410–7425.

3 Greer EL, Brunet A: Signaling networks in aging. J Cell Sci 2008;121:407–412.

4 Greer EL, Dowlatshahi D, Banko MR, Villen J, Hoang K, Blanchard D, Gygi SP, Brunet A: An AMPK-FOXO pathway mediates longevity induced by a novel method of dietary restriction in *C. elegans*. Curr Biol 2007;17:1646–1656.

5 Alexander A, Cai SL, Kim J, Nanez A, Sahin M, MacLean KH, Inoki K, Guan KL, Shen J, Person MD, et al: ATM signals to TSC2 in the cytoplasm to regulate mTORC1 in response to ROS. Proc Natl Acad Sci USA 2010;107:4153–4158.

6 Blagosklonny MV: Aging: ROS or TOR. Cell Cycle 2008;7:3344–3354.

7 Wullschleger S, Loewith R, Hall MN: TOR signaling in growth and metabolism. Cell 2006;124:471–484.

8 Lee JH, Budanov AV, Park EJ, Birse R, Kim TE, Perkins GA, Ocorr K, Ellisman MH, Bodmer R, Bier E, et al: Sestrin as a feedback inhibitor of TOR that prevents age-related pathologies. Science 2010;327:1223–1228.

9 Luong N, Davies CR, Wessells RJ, Graham SM, King MT, Veech R, Bodmer R, Oldham SM: Activated FOXO-mediated insulin resistance is blocked by reduction of TOR activity. Cell Metab 2006;4:133–142.

10 Wessells R, Fitzgerald E, Piazza N, Ocorr K, Morley S, Davies C, Lim HY, Elmen L, Hayes M, Oldham S, et al: d4eBP acts downstream of both dTOR and dFoxo to modulate cardiac functional aging in *Drosophila*. Aging Cell 2009;8:542–552.

11 Wessells RJ, Fitzgerald E, Cypser JR, Tatar M, Bodmer R: Insulin regulation of heart function in aging fruit flies. Nat Genet 2004;36:1275–1281.

12 Lee JH, Budanov AV, Talukdar S, Park EJ, Park HL, Park HW, Bandyopadhyay G, Li N, Aghajan M, Jang I, et al: Maintenance of metabolic homeostasis by Sestrin2 and Sestrin3. Cell Metab 2012;16:311–321.

13 Budanov AV, Lee JH, Karin M: Stressin' Sestrins take an aging fight. EMBO Mol Med 2010;2:388–400.

Duozhuang Tang, MD
Leibniz Institute for Age Research, Fritz Lipmann Institute e.V. (FLI)
Beutenbergstrasse 11
DE–07745 Jena (Germany)
E-Mail dtang@fli-leibniz.de

Rudolph KL (ed): Adult Stem Cells in Aging, Diseases and Cancer.
Else Kröner-Fresenius Symp. Basel, Karger, 2015, vol 5, pp 69–73 (DOI: 10.1159/000366567)

Telomeres and Stem Cell Aging

Daniel A. Felix

Leibniz Institute for Age Research, Fritz Lipmann Institute, Jena, Germany

Abstract

Telomeres are specialized regions at the ends of each chromosome that protect them from erosion and the erroneous recognition as DNA breaks. As telomeres shorten with each cell division, the enzyme telomerase is needed to maintain a functional and healthy telomere length. However, only germ and stem cells retain a sufficient telomerase activity to do so. In the absence of telomerase, the telomeres get critically short which induces senescence of the cell. This mechanism limits the number of times a cell can divide and so prevents runaway proliferation and cancer, but by doing so this mechanism is also directly responsible for aging. The interplay between telomeres, telomerase, DNA damage response and senescence as well as their relation to cancer and aging makes this a dynamic and fruitful area of research with high hopes for medical advances.

A DNA double-strand break is a very serious accident that splits a chromosome in two and will lead to the loss of genetic information if not corrected. Fortunately, our cells have repair mechanisms which are able to find the two ends of the break and splice them together again.

At a first glance, the ends of the chromosomes look very much the same as a DNA break; however, they are protected by a specialized structure called telomeres, which are formed by repeated DNA sequences and specialized binding proteins that together form a dynamic protective cap. Without this protective cap, the chromosome ends would induce the erroneous activation of DNA repair mechanisms which would result in deleterious homologous recombination, end-to-end fusions, genome rearrangements and ultimately death of the cell or, conversely, cancerous growth [1].

Because of the mechanism of DNA replication, telomeres shorten with each cell division. This imposes an upper limit on the number of times a cell can divide, called the Hayflick limit, which lies between 50 and 70 divisions for a human cell [2]. Critically short telomeres lose their protective function and are recognized by a host of DNA damage-recognizing proteins, most importantly ATM and ATR, which will form telomere dysfunction-induced foci resulting in p53-dependent senescence, effectively shutting down the cell [3]. This mechanism has been recognized as an important bulwark against the runaway proliferation of cancer, but is at the same time one of the most important contributors to aging [4].

In order for a cell to overcome the Hayflick limit, it needs to elongate its telomeres. The most commonly used telomere maintenance process is the addition of new telomeric repeats by an enzymatic complex called telomerase reverse transcriptase (telomerase). Active telomerase in adults is found mainly in germ cells and stem cells, but is nearly absent in differentiated somatic tissues. Artificial expression of telomerase in many cell types has been shown to remove the Hayflick limit, allowing a near unlimited proliferation capacity [5]. So, it is not surprising that telomerase activity has been found to be significantly increased in 80–90% of all cancers [6].

Alternatively, tumors without telomerase activity often exhibit activation of the alternative lengthening of telomeres (ALT) mechanism which relies on recombination-mediated elongation [7]. ALT is not normally found in vertebrate cells but is an aberration which abuses normal DNA repair mechanisms and needs several alterations in the cell before becoming active.

In 2009, the Nobel Prize in Physiology or Medicine was awarded to Drs. Elizabeth H. Blackburn, Jack W. Szostak and Carol W. Greider for their discovery of how chromosome ends are protected by telomeres and replicated by the enzyme telomerase [8, 9]. We have come a long way since their groundbreaking discoveries, but there is still much to learn about telomere biology, and the recent advances in relation to aging have been discussed at the 5th Else Kröner-Fresenius Symposium.

Telomerase, Adult Stem Cells and Cancer

Telomere length plays a dual role in cancer biology. On one hand, shortened telomeres in somatic cells, which are generally telomerase free, will induce p53-mediated senescence and present therefore an excellent stopgap measure to unlimited proliferation of mutant (pre-)cancerous cell clones. On the other hand, shortened telomeres induce chromosomal instability, which, together with subsequent reactivation of telomerase, can promote cancer formation.

But as Dr. Lenhard K. Rudolph explained, the picture is less clear when we look at adult somatic stem and progenitor cells that maintain a low level of active telomerase but still suffer some telomere shortening during each cell division. The slower erosion of their telomeres probably enables those cells to divide with near-critical telomere length more often than telomerase-free differentiated cells, thus increasing their probability of acquiring chromosomal instability and accumulation of pro-carcinogenic mutations [10]. In addition, it might be easier for a cell to just increase a preexisting telomerase expression compared to a de novo activation of telomerase in differentiated somatic cells.

In the light of the growing evidence about the role of adult stem cells in the origin of cancer formation [11], Dr. Lenhard K. Rudolph then highlighted the importance to analyze the interplay of telomerase expression and the molecular mechanisms that allow the clonal evolution towards a cancer cell.

Studying Telomerase in vivo

Telomerase is also emerging as a crucial stem cell factor. Telomerase knockout mice display a host of stem cell-related problems that range from infertility, impaired hematopoietic stem cell (HSC) self-renewal, intestinal dysfunction and premature aging syndrome [12]. Conversely, induced expression of TERT in senescent hair follicles is able to reactivate dormant stem cells leading to the regrowth of hair [13].

But, as Steven Artandi explained, while telomerase is clearly relevant for stem cell diseases, telomerase research has been clearly hampered by the low expression levels of its components TERT and TERC. This led to conflicting reports on the expression levels of telomerase, especially in mouse, and no good tools exist to isolate telomerase-positive cells for subsequent study. To circumvent this problem his lab has genetically engineered mice to express a fluorescent protein in cells that activate the telomerase gene. So his lab is now able to identify cells which activate telomerase by their fluorescence and then follow the process of differentiation, which, as predicted, leads to a loss of telomerase expression as the cells differentiate and consequently those cells also lose their diagnostic fluorescence. Furthermore, thanks to these transgenic mice it is now possible for the first time to isolate highly pure populations of telomerase-active stem cells for further study.

Stalled Replication Induces the ALT Pathway

Telomere shortening acts as a tumor-suppressive mechanism by limiting the proliferative capacity of dividing cells. Therefore, telomerase is generally only active in the germ line and stem cells. However, a cell that obtains the ability to maintain its telomeres is rendered immortal and has acquired one of the hallmarks of cancer. Most often, this will happen by alterations that will activate telomerase in a cell, but in 5–10% of human cancers the ALT pathway is activated, which relies not on telomerase but on recombination-mediated elongation. Since the ALT pathway can be activated by the inhibition of telomerase [12], it is important to understand the mechanism of activation of the ALT pathway if we want to target telomere length as a therapy, urged Dr. Jan Karlseder. He went on to explain that chromatin, the packaging state of the DNA, had recently been linked to the ALT pathway. The chromatin maintenance genes ATRX, DAXX or histone H3.3 had been found to be very frequently mutated in cancers that use ALT [14–16] and the histone marks at telomeres are different in cells that use ALT to the ones using telomerase [17]. He has focused his work on the histone Chaperone ASF1, a protein that removes the histones in front of the advancing DNA replication fork and puts histones back into the DNA after it has been replicated. If ASF1 is missing, the replisome will stall and RPA protein will bind to the free single-stranded DNA. As a result, the newly synthesized DNA cannot be packed onto histones. His lab has demonstrated that ASF1 silencing induces colocalization of RPA2 and PML specifically in large foci at the telomere ends the same way as it happens in telomere recombination centers of ALT cells. By tagging the telomeres of the leading strand in red and of the lagging strand in green, this recombination event in ASF1 silenced cells was demonstrated directly since the resulting telomere ends were now labeled yellow.

Importantly, even after removal of siASF1 and the resulting reexpression of ASF1 protein, the cells remained permanently in ALT mode, indicating an epigenetic maintenance of the ALT pathway activation [18]. Interestingly, transcriptional profiling indicated a similar response to the ASF1 knockdown as to an inflammatory event or viral infection, which could have medicinal relevance regarding the treatment of ALT cells.

Breast Stem Cells with High Telomerase Activity but Short Telomeres

Emerging evidence suggests that adult tissue-specific stem cells are key to the development and cancer of the human breast. In a hierarchical manner, these divide asymmetrically, producing an increasing number of cells with distinct properties [19]. Understanding the biological characteristics of human breast stem cells and their progeny is crucial in attempts to understand and fight breast cancer.

Development and homeostasis of the human breast is thought to be a product of a hierarchical cascade of stem cells giving rise to the two main mammary cell lineages with luminal progenitors and myoepithelial progenitors being transiently amplifying progenitor populations. But as Connie Eaves reported at the Else Kröner-Fresenius Symposium, the luminal progenitor cells have greatly shortened telomeres, and as a result show an increased activation of the DNA damage response genes MRE11, RAD50, ATM, ATR and BLM and their localization at telomeres. Surprisingly, however, she found that luminal progenitor cells have a very high telomerase activity that decreases with age [20]. These findings question the strictly hierarchical model of tissue homeostasis in the human breast and reveal features of the luminal progenitors that appear to make them prime candidates for the origin of breast cancer.

Pot1a Expression Enhances Colony Formation Ability of Hematopoietic Stem Cells

Hematopoiesis depends on the controlled self-renewal of the HSCs in the bone marrow. As we age, accumulated DNA damage decreases the bone marrow long-term repopulation ability of HSCs. Protection of telomeres 1 (Pot1) is a crucial component of the shelterin complex which is important for telomere maintenance preventing the chromosome ends from erroneously inducing DNA damage signaling [21]. Previously, the Lab of Toshio Suda had found that Pot1a is highly expressed in HSCs but that expression decreased as the mouse aged or during in vitro culture of the HSCs.

As he now reported at the Else Kröner-Fresenius Symposium, knocking down Pot1a function in HSCs increased expression of the telomeric DNA damage response genes Trf1 and 53BP1 together with a highly increased rate of telomere dysfunction-induced foci. In addition, the cells appeared to become senescent since levels of p16 and p19 became elevated, while Ezh2 and Irf8 were downregulated. In accordance with these findings, shPot1a HSCs perform poorly in long-term repopulation activity when transplanted.

Conversely, overexpression of Pot1a in cultured HSC greatly enhanced their colony formation capacity in culture as well as their engraftment capacity in primary and secondary bone marrow transplants. Dr. Suda went on to explain that they observed repression of ATR signaling in Pot1a overexpressing HSCs, which would lead to reduced senescence of those cells. Moreover, the most primitive HSCs (LSK) expressing high levels of Pot1a were found to have very low levels of harmful reactive oxygen species (ROS) compared to low-expressing cells, and Pot1a overexpression was able to significantly decrease ROS.

Finally, his group was able to enhance engraftment and self-renewal capacity of HSCs by directly adding mouse or human Pot1a protein fused to a membrane-transferring motive to the culture medium, showing a potential way to enhance the success rate of bone marrow transplants [22].

Conclusions

Our knowledge about the interplay of telomere maintenance, DNA damage response and checkpoint-induced senescence is growing rapidly. This understanding is opening up new avenues for fighting aging and cancer, cultivating and maintaining healthy stem cells and for treating degenerative diseases. As it emerges that short telomeres by themselves are able to induce DNA damage, it will be important to address this problem in induced pluripotent stem cells in order to fully develop their therapeutic possibilities.

References

1 Kim NW, et al: Specific association of human telomerase activity with immortal cells and cancer. Science 1994; 266:2011–2015.
2 Hayflick L: The limited in vitro lifetime of human diploid cell strains. Exp Cell Res 1965;37:614–636.
3 Takai H, Smogorzewska A, de Lange T: DNA damage foci at dysfunctional telomeres. Curr Biol 2003;13:1549–1556.
4 Blasco MA et al: Telomere shortening and tumor formation by mouse cells lacking telomerase RNA. Cell 1997;91: 25–34.
5 Bryan TM, Reddel RR: Telomere dynamics and telomerase activity in in vitro immortalised human cells. Eur J Cancer 1997;33:767–773.
6 Shay JW, Bacchetti S: A survey of telomerase activity in human cancer. Eur J Cancer 1997;33:787–791.
7 Dunham MA, Neumann AA, Fasching CL, Reddel RR: Telomere maintenance by recombination in human cells. Nat Genet 2000;26:447–450.
8 Greider CW, Blackburn EH: The telomere terminal transferase of Tetrahymena is a ribonucleoprotein enzyme with two kinds of primer specificity. Cell 1987;51:887–898.
9 Harley C, Futcher A, Greider C: Telomeres shorten during ageing of human fibroblasts. Nature 1990;345:458–460.
10 Günes C, Rudolph KL: The role of telomeres in stem cells and cancer. Cell 2013;152:390–393.
11 Welch JS, et al: The origin and evolution of mutations in acute myeloid leukemia. Cell 2012;150:264–278.
12 Young NS: Telomere biology and telomere diseases: implications for practice and research. Hematology Am Soc Hematol Educ Program 2010;2010;30–35.
13 Jan H-M, et al: The use of polyethylenimine-DNA to topically deliver hTERT to promote hair growth. Gene Ther 2012;19:86–93.
14 Lovejoy CA, et al: Loss of ATRX, genome instability, and an altered DNA damage response are hallmarks of the alternative lengthening of telomeres pathway. PLoS Genet 2012;8:e1002772.
15 Bower K, et al: Loss of wild-type ATRX expression in somatic cell hybrids segregates with activation of alternative lengthening of telomeres. PLoS One 2012;7:e50062.
16 Lewis PW, Elsaesser SJ, Noh K-M, Stadler SC, Allis CD: Daxx is an H3.3-specific histone chaperone and cooperates with ATRX in replication-independent chromatin assembly at telomeres. Proc Natl Acad Sci USA 2010;107: 14075–14080.
17 Benetti R, et al: Suv4-20h deficiency results in telomere elongation and derepression of telomere recombination. J Cell Biol 2007;178:925–936.
18 O'Sullivan RJ, et al: Rapid induction of alternative lengthening of telomeres by depletion of the histone chaperone ASF1. Nat Struct Mol Biol 2014;21:167–174.
19 Petersen OW, Polyak K: Stem cells in the human breast. Cold Spring Harb Perspect Biol 2010;2:a003160.
20 Kannan N, et al: The luminal progenitor compartment of the normal human mammary gland constitutes a unique site of telomere dysfunction. Stem Cell Reports 2013;1:28–37.
21 Loayza D, De Lange T: POT1 as a terminal transducer of TRF1 telomere length control. Nature 2003;423:1013–1018.
22 Hosokawa K, Suda T, Arai F: Shelterin component protection of telomeres 1 (Pot1) plays a critical role in the self-renewal of HSCs. Exp Hematol 2013; 41:S37.

Daniel A. Felix, PhD
Leibniz Institute for Age Research, Fritz Lipmann Institute e.V. (FLI)
Beutenbergstrasse 11
DE–07745 Jena (Germany)
E-Mail dfelix@fli-leibniz.de

Rudolph KL (ed): Adult Stem Cells in Aging, Diseases and Cancer.
Else Kröner-Fresenius Symp. Basel, Karger, 2015, vol 5, pp 74–82 (DOI: 10.1159/000366568)

DNA Damage and Checkpoint Responses in Adult Stem Cells

Vasily Romanov[a] · Aruna Shukla[b] · Zhenyu Ju[b]

[a]Leibniz Institute for Age Research – Fritz Lipmann Institute, Jena, Germany; [b]Institute of Aging Research, School of Medicine, Hangzhou Normal University, Hangzhou, China

Abstract

All living cells, including tissue-specific stem cells, are constantly exposed to DNA damage, such as altered DNA bases and DNA double-strand breaks. Experimental evidence suggests that DNA damage checkpoint genes not only influence stem cell function in response to DNA damage but also influence the maintenance of undamaged stem cells by controlling self-renewal, differentiation and quiescence. Modulating the activity of DNA damage checkpoints can either accelerate or decelerate tissue aging and age-related carcinogenesis. The outcome depends on cell-intrinsic and cell-extrinsic mechanisms, the level of DNA damage itself, and possibly also on the fidelity of molecular mechanisms controlling the clearance of damaged cells from aging tissues. In this chapter, we review recent findings related to DNA damage and checkpoint responses in adult stem cells that were discussed at the 5th Else Kröner-Fresenius Symposium on Adult Stem Cells in Aging, Diseases and Cancer.

Aging is a complex biological process that increases the risk of a broad spectrum of human diseases and disorders [1–3]. The molecular mechanisms underlying the progressive physiologic decline with aging remain incompletely understood [4]. Accumulating evidence suggests that genome instability increases with age [5–7]. A number of potential mechanisms, often overlapping, have been proposed to explain age-dependent genome instability. These include the accumulation of oxidative damage to DNA, defects in mitochondrial function that promote oxidative damage to DNA and other macromolecules, mutations in proteins required for efficient DNA replication, DNA repair and checkpoints, telomere erosion and epigenetic effects on DNA repair and other genome maintenance programs [5–9].

The performance of adult stem cells is crucial for tissue homeostasis but their regenerative capacity declines with age, which is thought to contribute to declines in organ function during aging [10, 11]. Quiescent hematopoietic stem cells (HSCs) accumulate more DNA damage than highly proliferative progenitor cells during aging, suggesting that stem cells tolerate higher levels of DNA damage than progenitor cells [12]. The resistance of stem cells to DNA damage is probably essential for organismal survival and tissue regeneration [13, 14]. DNA damage checkpoints and DNA repair pathways, which can either repair the damage or lead to senescence, apoptosis and differentiation to eliminate the damaged stem cells, are often activated in a cell cycle-dependent manner [12, 14, 15]. It remains yet to be explored whether these cell cycle-dependent regulations of checkpoints and repair pathways are the underlying mechanism of DNA damage accumulation in quiescent HSCs compared to cycling progenitor cells.

DNA Damage Accumulation during Hematopoietic Stem Cell Aging

The primary function of tissue-specific stem cells in the adult organism is to mediate tissue and organ homeostasis and regenerate tissue after injury. Regardless of the system being studied, aging is accompanied by reduced regenerative potential to maintain tissue homeostasis [10]. Dr. Derrick Rossi and colleagues used HSCs as a model system to understand the accumulation of DNA damage during aging and its physiological impact on stem cell function. Using genetically engineered mice deficient in several genomic maintenance pathways including nucleotide excision repair, telomere maintenance and nonhomologous end-joining, Dr. Rossi and colleagues showed that although deficiencies in these pathways did not deplete stem cell reserves with age, stem cell functional capacity was severely affected under conditions of genotoxic stress. Moreover, they provided evidence that endogenous DNA damage accumulates with age in wild-type stem cells, which represents a physiological mechanism of stem cell aging that may contribute to the diminished capacity of aged tissues to maintain homeostasis after exposure to acute stress or injury [16]. So, aging may be at least in part attributable to progressive stem cell dysfunction as a consequence of DNA damage accumulation. It remains yet to be delineated to what extent aging-related alterations in the hematopoietic system are attributable to aging of the stem cell compartment itself. Another important question is to delineate mechanisms that are responsible for the accumulation of DNA damage in quiescent HSCs compared to cycling progenitor cells. During the 5th Else Kröner-Fresenius Symposium, Dr. Rossi presented new data analyzing the role of HSC quiescence in affecting checkpoint response and DNA repair. For these studies, he teamed up with Dr. Seita, who focused over recent years on gene expression analysis in highly purified subpopulations of hematopoietic stem and progenitor cells [17]. Dr. Rossi and colleagues performed gene expression profiling of DNA repair and checkpoint responses in HSCs isolated from young (3–4 months; yHSCs) and old (20–25 months; oHSCs) mice versus multipotent progenitor cells (MPPs). The result showed that base excision repair (BER), homologous recombination (HR), mismatch repair (MMR) and checkpoint pathways were significantly attenuated in both yHSCs and oHSCs in comparison to MPPs. Further, the analysis of cell cycle status of yHSCs and oHSCs versus fetal HSCs (fHSCs) demonstrated that whereas the cell cycle profiles of yHSCs and oHSCs were comparable, the percentage of proliferating HSCs in fetal liver was higher. Gene expression profiling of BER, MMR, HR and checkpoint pathways revealed the decreasing pattern of gene expression from fHSCs to yHSCs and oHSCs, which led to the conclusion that DNA repair and response pathways were attenuated in quiescent HSCs but not in cycling HSCs. To find out whether oHSCs have an attenuated ability to repair the DNA damage once they enter into cell cycle, Dr. Rossi and colleagues sorted yHSCs and oHSCs, cultured them for 3, 6, 12 or 24 h, and then performed gene expression array. Although both yHSCs and oHSCs upregulated DNA repair pathways upon entry into the cell cycle, oHSCs exhibited delayed cell cycle entry kinetics compared to yHSCs, which showed a relatively high expression level of cell cycle-specific cyclins at these four time points. Interestingly, an exit of HSCs from G0 was followed by strong upregulation of *CDKN1A* and *GADD45* genes marking a p21/Gadd45-mediated DNA damage response. Dr. Rossi and colleagues also analyzed repair of DNA damage in oHSCs versus yHSCs upon cell cycle entry and found that both of them effectively repair DNA damage in 24 h after cell cycle induction. Importantly, they also showed that yHSCs and oHSCs were both replication competent and exhibited comparable clonal activity.

p53-/p21-Dependent DNA Damage Response in Stem Cells

Since the accumulation of DNA damage is a unique problem for HSCs, the next question is how the HSCs deal with it. During the meeting, Dr. Pier Giuseppe Pelicci showed interesting evidence that HSCs have developed a distinct mechanism of DNA damage checkpoint responses compared to the progenitor cells, for example in response to oncogene-induced DNA damage.

It is known that expression of oncogenes like Ras, B-Raf, or Myc in MEFs induces DNA damage and activates a p53-dependent checkpoint response leading to senescence or apoptosis [18–22]. In the absence of p53, oncogene expression induces cancer transformation [23]. Dr. Pelicci and his colleagues showed that the expression of leukemia-associated oncogenes such as PML-RAR and AML1-ETO fusion proteins in fibroblasts leads to p53-independent but p21-dependent senescence. However, expression of PML-RAR and AML1-ETO in normal HSCs induces DNA damage leading to activation of p21 and extended stem cell renewal [24]. In the absence of p21, such oncogene-expressing pre-leukemia stem cells functionally exhaust, and leukemogenesis does not proceed. These results suggested that oncogenes can induce opposite DNA damage responses depending on differentiation status and self-renewal capacity of the cells. A possible explanation suggests that there are differences in the processing of DNA damage and/or the induction of DNA damage signals in stem cells compared to differentiated cells.

During the Symposium, Dr. Pelicci presented new data from his group on p53-/p21-dependent DNA damage response in HSCs and mammary stem cells (MaSCs) [25]. To investigate HSCs after DNA damage, Dr. Pelicci and his colleagues irradiated mice (wild-type, $p53^{-/-}$ or $p21^{-/-}$) by X-rays using the maximal sublethal dose of 5.5 Gy, and isolated HSCs from bone marrow. The experiments revealed that irradiation (IR) of wild-type mice induced robust accumulation of both phospho-p53 and p21 in the HSC progeny, MPPs and common myeloid progenitors (CMPs), and also led to strong induction of apoptosis characterized by upregulation of cleaved caspase-3. Apoptosis of MPPs and CMPs was strictly p53 dependent and was rescued in p53-null mice, but did not depend on the p21 upregulation since $p21^{-/-}$ mice showed no rescue.

In contrast to the data on HSC progeny, HSCs themselves showed only a marginal accumulation of activated phospho-p53 and no induction of apoptosis in response to IR. γH2AX staining of DNA damage foci revealed that the apoptosis resistance of irradiated HSCs was not associated with reduced levels of DNA damage. It also appeared to be independent of HSC quiescence since apoptosis resistance was observed in both quiescent HSCs and proliferating HSCs that were stimulated to divide by G-CSF treatment.

Dr. Pelicci and colleagues observed an accumulation of p21-positive HSCs in response to IR, which was independent of p53 activation. Deletion of p53 did not reduce the accumulation of p21-positive cells in irradiated HSCs. Instead, the study revealed that the induction of p21 had suppressed p53 activation. In contrast to wild-type HSCs, $p21^{-/-}$ HSCs showed an increased accumulation of phospho-p53-positive cells after IR, which was associated with activation of the apoptotic program. This p21-dependent suppression of p53 induction in response to IR was not observed at progenitor cell level.

Next, Dr. Pelicci and colleagues addressed the question whether p21 upregulation modulates DNA damage response in irradiated HSCs only by inhibiting p53 and apoptosis or whether p21 also activates cellular pathways involved in DNA repair or maintenance of self-renewal of stem cells after DNA damage. The study revealed evidence that wild-type and $p53^{-/-}$ HSCs showed γH2AX staining 6 h after IR, which disappeared almost completely 2 months later. In contrast, approximately one third of $p21^{-/-}$ HSCs still retained γH2AX staining 2 months after IR, suggesting that p21 contributes to DNA repair in HSCs. In addition, competitive bone marrow transplantation experiment showed that the persistent DNA damage in $p21^{-/-}$ HSCs 2 months after IR reduced their repopulating ability. Together, these data demonstrated that p21 activates cellular pathways that limit accumulation of DNA damage and self-renewal defects of stem cells in the context of DNA damage.

In order to understand the mechanism through which p21, a well-known negative regulator of cell cycle, contributes to DNA damage response in stem cells, Dr. Pelicci and colleagues investigated the cell proliferation fate of stem cells after IR. Wild-type HSCs exhibited enhanced proliferation (BrdU, Ki67) and increased cell number after IR. Surprisingly, the deletion of p21 suppressed the proliferation and reduced IR-induced HSC expansion. These findings indicate that the upregulation of p21 upon IR not only prevents p53 activation and apoptosis, but also activates DNA repair, maintains HSC self-renewal, and induces expansion of the HSC pool by activating HSC proliferation.

Using another stem cell model, MaSCs, Dr. Pelicci and colleagues investigated how p21 controls the increasing of stem cell number in response to IR. It is known that the stem cell number is regulated by the modality of stem cell divisions. While symmetric self-renewal divisions of one stem cell to two stem cells increase their number, asymmetric divisions of one stem cell to one stem cell and one progenitor cell result in a constant number of stem cells, and symmetric differentiation divisions of one stem cell into two progenitor cells lead to a reduction of stem cell number. The fact that the number of stem cells increased in response to IR suggested that IR could switch the modality of divisions, from asymmetric to symmetric self-renewal divisions.

Dr. Pelicci and colleagues previously reported that, in contrast to wild-type MaSCs, $p53^{-/-}$ MaSCs mainly divide symmetrically, indicating that the modality of stem cell divisions is regulated by p53 [26, 27]. At the 5th Else Kröner-Fresenius Symposium, Dr. Pelicci presented new data showing that IR induces symmetric divisions in a p21-dependent manner. In addition, by using a system with the expression of GFP under p53-dependent promoter that allows in vivo fluorescence-tracing of p53 activity in single cell, he showed that the asymmetric division (characterized by the increasing level of p53 activity) of MaSC was switched to symmetric division (characterized by the decreasing of p53 activity) of stem cells by IR. And, interestingly, both down-regulation of p53 activity and symmetric division of irradiated stem cells were strongly depending on the presence of p21.

Dr. Pelicci's talk emphasized a unique DNA damage checkpoint response in adult stem cells compared to their progeny. While DNA damage leads to p53 activation and subsequent senescence or apoptosis responses to suppress tumor formation in progenitor cells, the same DNA damage causes p21 accumulation, which potentiates DNA repair and leads to symmetric division in stem cells (fig. 1). The logical question then would be how transformation is suppressed in stem cells. Indeed, Dr. Pelicci showed that DNA repair was never complete, and self-renewal was reduced in stem cells recovering from DNA dam-

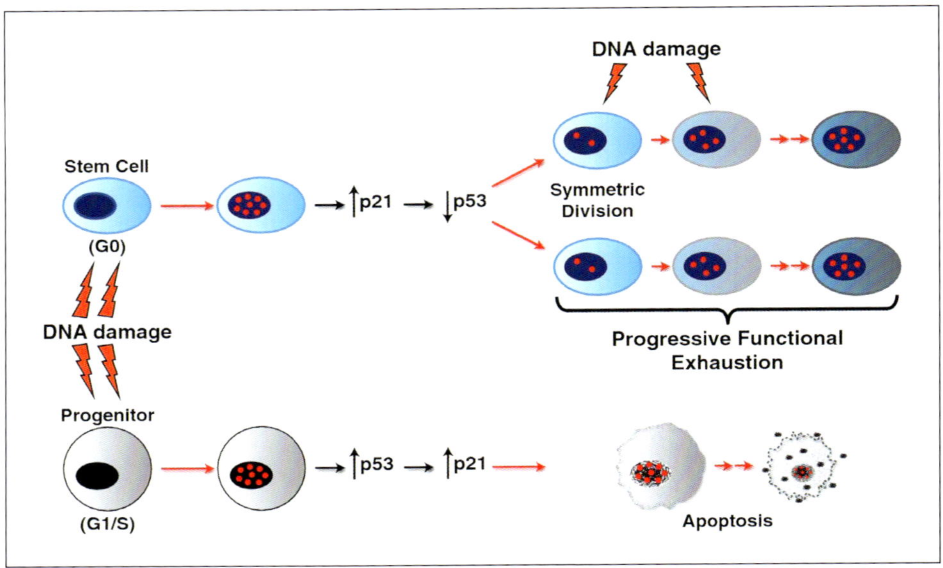

Fig. 1. The model suggested by Dr. Pelicci and colleagues: effect of X-rays on stem cells (SCs) and progenitors. IR of SCs upregulates p21, which inhibits p53 basal activity and prevents p53 activation, impeding apoptosis and leading to cell cycle entry and symmetric self-renewing divisions. p21 also activates DNA repair in SCs, limiting DNA damage accrual and self-renewal exhaustion. DNA damage repair is never complete, however, and in response to continuous exposure to DNA-damaging events, SCs progressively accumulate persistent DNA damage and become functionally exhausted. In contrast, progenitor cells respond to X-rays by activating a checkpoint response that involves p53 and its target p21, leading to apoptosis. Reprinted from Insinga et al. [25].

age. Thus, he concluded that the DNA repair activity is 'flawed' in stem cells allowing accumulation of persistent DNA damage and thus limiting their self-renewal potential. Taken together, Dr. Pelicci's findings imply that such a unique p21-dependent DNA damage response has evolved in stem cells to allow rapid recovery from acute damage and to prevent immortalization by limiting their long-term survival.

DNA Damage Response in Human Hematopoietic Stem Cells

The above data from mouse models clearly demonstrated that DNA damage response is very crucial in the regulation of HSC homeostasis in response to DNA damage. At the 5th Else Kröner-Fresenius Symposium, Dr. Michael Milyavsky showed interesting data of differences in DNA damage response between self-renewing human HSCs and differentiated cells, which lack self-renewal capacity.

Dr. Milyavsky and colleagues characterized cell death response of umbilical cord blood HSCs and their progeny, MPPs and committed progenitors (CPs), after sublethal doses of IR. Both HSC and MPP fractions showed significantly higher proportion of apoptosis than CPs [28]. Further experiments revealed that human HSCs exhibit delayed DNA double-strand break (DSB) rejoining, persistent γH2AX foci, and enhanced p53- and ASPP1-dependent apoptosis after gamma IR compared to progenitors. Interestingly, the group of Emmanuelle Passegue found that murine HSCs have unique cell-intrinsic mechanisms ensuring their survival in response to ionizing IR, which include enhanced pro-survival gene expression and strong activation of p53-mediated DNA damage response. Especially, they showed that quiescent and proliferating HSCs are equally better radioprotected than more differentiated cells, although they use different types of DNA repair mechanisms [15]. Together, these results indicate that self-renewing HSCs have distinct mechanism from their progeny to deal with DNA damage and also that mouse and human HSCs are different in their DNA damage repair response. The group of Dr. Daohong Zhou directly proved this conclusion and reported that human HSCs are less proficient in the repair of DSBs than CPs, whereas mouse HSCs repaired the damage more efficiently than CPs. The difference in the abilities of human and mouse HSCs and CPs to repair DSBs is likely attributed to their differential expression of key DNA damage repair genes such as *LIG4* [29].

To determine whether p53 or Bcl-2 modulation affects human HSC self-renewal, Dr. Milyavsky and colleagues used NOD/SCID repopulation assay: human HSCs were transduced with control, Bcl-2 or dominant-negative p53KD-expressing lentiviruses, then exposed to gamma IR, followed by transplantation into immune-deficient NOD/SCID mice. In nonirradiated HSCs, only Bcl-2 overexpression resulted in significantly higher self-renewal ability. The p53KD group showed a trend to lower engraftment compared to the Bcl-2 group. To provide possible explanation for the differential effect of p53 inactivation and Bcl-2 overexpression on self-renewal, the extent of spontaneous DNA damage was examined in HSC progeny. The results showed that human Lin^- cells from primary recipients of control, p53KD- and Bcl-2-overexpressing cells harbored approximately the same number of γH2AX foci as human fibroblasts. However, upon serial transplantation, there was a striking elevation in the number of γH2AX foci per cell only in human Lin^- cells derived from HSCs with disabled p53 as compared to Lin^- cells derived from control HSCs or Bcl-2-overexpressing HSCs. These results indicate that the increase in γH2AX foci seen solely in p53KD cells is likely due to a loss of one or more p53 functions and must be independent of apoptosis since abrogation of apoptosis induction by Bcl-2 overexpression did not result in an elevation in foci numbers. Together, these data indicated that p53 has some role in regulating self-renewal of human HSCs in an apoptosis-independent manner.

Based on these findings, Dr. Milyavsky and colleagues went on to identify regulators of the DNA damage response in hematopoietic cells by combining genome-wide shRNA screening with gene expression profiling and using in vitro assay and functional transplantation model. For this, they utilized TEX hematopoietic cell line, which is very sensitive to DNA damage, infected with genome-wide shRNA library, and then exposed repeatedly to several rounds of IR. The hypothesis was that any shRNA that gives a protective effect to the cell line will be enriched in the genomic DNA after serial IR treatment. The results showed that a number of crucial genes, including *TP53*, *CHEK2* and *SMYD2*, were identified. But very interestingly, the majority of the candidates which make HSCs resistant to IR were not previously connected with DNA damage response. Considering that some of shRNA hits from the screen might be markers of therapy sensitivity in primary human acute myeloid leukemia (AML), they compared their list with the results on genetic screen of AMLs resistant to etoposide-induced DNA damage (in collaboration with Dr. Sean McDermott). Interestingly, several candidates that they found indeed demonstrated higher expression in etoposide-sensitive samples, suggesting that at least part of the genes from performed shRNA screen could also serve as novel markers of sensitivity/resistance of AMLs to DNA damage. The specific role of these genes in the biology of AML and normal HSCs remains to be defined.

Stem and Progenitor Cell Senescence in Aging

The ultimate consequence of DNA damage response is cellular senescence or apoptosis. Whether stem cells enter cellular senescence in response to genotoxic stress is currently not well characterized. Also, it remains to be delineated whether stem cell senescence contributes to organismal aging. During the 5th Else Kröner-Fresenius Symposium, Dr. Jan van Deursen presented very interesting evidence that aneuploidy-induced cellular senescence can influence tissue aging.

Normally, deletion of mitotic checkpoints leads to cell lethality due to the induction of mitotic catastrophe. To circumvent this problem and to make possible the analysis of aneuploidy at an organismal level, Dr. van Deursen and colleagues developed mouse mutants with reduced activity of ploidy control genes, but not complete knockout of the genes. One of these mouse models carries a mutation with very low expression (10% of normal level) of the mitotic checkpoint protein BubR1, which is one of the key components ensuring proper chromosome segregation [30]. Mutations in human *BubR1* are associated with mosaic variegated aneuploidy, a rare syndrome that is characterized by aneuploidization and various progeroid traits, including short lifespan, short stature, cataracts, facial dysmorphisms, and mental retardation [31–33]. These BubR1 mutant mice did not exhibit an increased rate of cancer formation but at the age of 3–4 months developed premature aging phenotype revealing cataracts, kyphosis, sarcopenia, and fat loss [30, 34–36]. The tissues affected by these progeroid phenotypes showed increased level of age-associated β-Gal staining and elevated p16 expression [37]. These results suggested that the induction of p16 and accumulation of senescent cells contributed to the development of age-associated phenotypes in this mouse model [37, 38]. Experiments from Dr. van Deursen's lab suggest that hyperactivity of APC/C^{cdc20} (an E3 ubiquitin ligase that drives mitosis) contributed to the induction of p16, activation of senescence program, and finally to the observed age-associated mouse phenotype.

Although the accumulation of senescent cells is implicated in contributing to aging and age-related pathologies, experimental proof for this concept has been lacking. What we do know is that senescent cells accumulate with aging in a variety of tissues, in a wide range of species including primates [39, 40]. Senescent cells were shown to secrete increased levels of cytokines, chemokines, proteinases and growth factors [41]. This phenotype is also known as the senescence-associated secretory phenotype (SASP). The SASP, which influences surrounding cells, possibly contributes to impaired homeostasis and functionality of aging tissues that accumulate senescent cells [41]. To directly test whether the accumulation of p16-positive senescent cells contributes to tissue aging, Dr. van Deursen's group used two approaches. The first was inactivation of p16 in BubR1-hypomorphic mice, and the second approach was clearance of senescent cells by using a transgene that contains p16 promoter and FKBP fusion protein that kills the cells when expressed upon the administration of dimerizer AP20187. Both approaches resulted in an attenuation of tissue deterioration, suggesting a causal relationship between accumulation of senescent cells and age-associated phenotypes [34, 42, 43].

But what type of cells undergoes senescence in the tissue? Dr. van Deursen and colleagues studied skeletal muscle and fat from a BubR1 progeroid model. For the skeletal muscle, different kinds of mononuclear cells like quiescent satellite cells (Q-SCs), activated satellite cells (A-SCs), fibro-/adipogenic progenitors (FAPs), and endothelial cells (ECs) were isolated using various sorting protocols. The majority of p16 expression was found in FAP cells, and a substantially significant expression was found in cell populations that contained A-SCs, whereas Q-SCs were completely negative for p16, as were myofibers and ECs. However, the number of Q-SCs and FAPs was not reduced in BubR1 mice. To investigate whether senescent progenitors impair tissue regeneration, Dr. van Deursen's group performed cyclophosphamide (CTX) injection experiments on the wild-type and the BubR1 progeroid model. After 18 days of CTX injection, BubR1 progeroid mice showed delayed tissue regeneration, more fibrosis, smaller and less myofibers, but deletion of p16 significantly improved tissue regeneration in BubR1 mutant mice. These results clearly indicated that senescent cells impair tissue regeneration. Similar results were obtained in fat tissues. Altogether, Dr. van Deursen's data suggested that the senescent progenitors contribute to the impaired tissue regeneration by producing age-associated secretome that affects the nearby cells, and clearance of the p16-positive senescent progenitor cells can improve the tissue regeneration. Along this line, the group of Prof. Dr. Karl Lenhard Rudolph showed genetic evidence that p53-dependent mechanisms prevent tissue destruction in response to telomere dysfunction by depleting genetically unstable intestinal stem cells in aging telomere-dysfunctional mice [44].

Outlook

The primary function of tissue-specific stem cells in the adult organism is to mediate tissue and organ homeostasis and regenerate tissue after injury. Regardless of the system being studied, aging is accompanied by reduced regenerative potential to maintain tissue homeostasis. Adult stem cells possess extensive self-renewal capacity in order to sustain the life-long tissue replenishment, and this also makes them good targets for cancer transformation. To maintain a long-term regeneration with protection from cancer transformation, adult stem cells must have adopted a unique DNA damage response machinery. Further investigation into the underlying mechanism of DNA damage responses in adult stem cells during aging and upon various stresses is important for the clinical applications aiming to rejuvenate the old stem cells and to treat age-related diseases.

References

1 Campisi J: Senescent cells, tumor suppression, and organismal aging: good citizens, bad neighbors. Cell 2005;120:513–522.

2 Lotz MK, Carames B: Autophagy and cartilage homeostasis mechanisms in joint health, aging and OA. Nat Rev Rheumatol 2011;7:579–587.

3 Blasco MA: Telomeres and human disease: ageing, cancer and beyond. Nat Rev Genet 2005;6:611–622.

4 Sahin E, DePinho RA: Axis of ageing: telomeres, p53 and mitochondria. Nat Rev Mol Cell Biol 2012;13:397–404.

5 Burhans WC, Weinberger M: DNA replication stress, genome instability and aging. Nucleic Acids Res 2007;35:7545–7556.

6 Vijg J, Suh Y: Genome instability and aging. Annu Rev Physiol 2013;75:645–668.

7 Lopez-Otin C, Blasco MA, Partridge L, Serrano M, Kroemer G: The hallmarks of aging. Cell 2013;153:1194–1217.

8 Newgard CB, Sharpless NE: Coming of age: molecular drivers of aging and therapeutic opportunities. J Clin Invest 2013;123:946–950.

9 Bratic A, Larsson NG: The role of mitochondria in aging. J Clin Invest 2013;123:951–957.

10 Rossi DJ, Jamieson CH, Weissman IL: Stems cells and the pathways to aging and cancer. Cell 2008;132:681–696.

11 Sperka T, Wang J, Rudolph KL: DNA damage checkpoints in stem cells, ageing and cancer. Nat Rev Mol Cell Biol 2012;13:579–590.

12 Rossi DJ, Seita J, Czechowicz A, Bhattacharya D, Bryder D, et al: Hematopoietic stem cell quiescence attenuates DNA damage response and permits DNA damage accumulation during aging. Cell Cycle 2007;6:2371–2376.

13 Lane AA, Scadden DT: Stem cells and DNA damage: persist or perish? Cell 2010;142:360–362.

14 Blanpain C, Mohrin M, Sotiropoulou PA, Passegue E: DNA-damage response in tissue-specific and cancer stem cells. Cell Stem Cell 2011;8:16–29.

15 Mohrin M, Bourke E, Alexander D, Warr MR, Barry-Holson K, et al: Hematopoietic stem cell quiescence promotes error-prone DNA repair and mutagenesis. Cell Stem Cell 2010;7:174–185.

16 Rossi DJ, Bryder D, Seita J, Nussenzweig A, Hoeijmakers J, et al: Deficiencies in DNA damage repair limit the function of haematopoietic stem cells with age. Nature 2007;447:725–729.
17 Seita J, Sahoo D, Rossi DJ, Bhattacharya D, Serwold T, et al: Gene Expression Commons: an open platform for absolute gene expression profiling. PLoS One 2012;7:e40321.
18 Di Micco R, Fumagalli M, Cicalese A, Piccinin S, Gasparini P, et al: Oncogene-induced senescence is a DNA damage response triggered by DNA hyper-replication. Nature 2006;444: 638–642.
19 Evan GI, Wyllie AH, Gilbert CS, Littlewood TD, Land H, et al: Induction of apoptosis in fibroblasts by c-myc protein. Cell 1992;69:119–128.
20 Serrano M, Lin AW, McCurrach ME, Beach D, Lowe SW: Oncogenic ras provokes premature cell senescence associated with accumulation of p53 and p16INK4a. Cell 1997;88:593–602.
21 Vafa O, Wade M, Kern S, Beeche M, Pandita TK, et al: c-Myc can induce DNA damage, increase reactive oxygen species, and mitigate p53 function: a mechanism for oncogene-induced genetic instability. Mol Cell 2002;9:1031–1044.
22 Zhu J, Woods D, McMahon M, Bishop JM: Senescence of human fibroblasts induced by oncogenic Raf. Genes Dev 1998;12:2997–3007.
23 Fukasawa K, Vande Woude GF: Synergy between the Mos/mitogen-activated protein kinase pathway and loss of p53 function in transformation and chromosome instability. Mol Cell Biol 1997;17:506–518.
24 Viale A, De Franco F, Orleth A, Cambiaghi V, Giuliani V, et al: Cell-cycle restriction limits DNA damage and maintains self-renewal of leukaemia stem cells. Nature 2009;457:51–56.
25 Insinga A, Cicalese A, Faretta M, Gallo B, Albano L, et al: DNA damage in stem cells activates p21, inhibits p53, and induces symmetric self-renewing divisions. Proc Natl Acad Sci USA 2013;110: 3931–3936.
26 Cicalese A, Bonizzi G, Pasi CE, Faretta M, Ronzoni S, et al: The tumor suppressor p53 regulates polarity of self-renewing divisions in mammary stem cells. Cell 2009;138:1083–1095.
27 Pece S, Tosoni D, Confalonieri S, Mazzarol G, Vecchi M, et al: Biological and molecular heterogeneity of breast cancers correlates with their cancer stem cell content. Cell 2010;140:62–73.
28 Milyavsky M, Gan OI, Trottier M, Komosa M, Tabach O, et al: A distinctive DNA damage response in human hematopoietic stem cells reveals an apoptosis-independent role for p53 in self-renewal. Cell Stem Cell 2010;7:186–197.
29 Shao L, Feng W, Lee KJ, Chen BP, Zhou D: A sensitive and quantitative polymerase chain reaction-based cell free in vitro non-homologous end joining assay for hematopoietic stem cells. PLoS One 2012;7:e33499.
30 Baker DJ, Jeganathan KB, Cameron JD, Thompson M, Juneja S, et al: BubR1 insufficiency causes early onset of aging-associated phenotypes and infertility in mice. Nat Genet 2004;36:744–749.
31 Hanks S, Coleman K, Reid S, Plaja A, Firth H, et al: Constitutional aneuploidy and cancer predisposition caused by biallelic mutations in BUB1B. Nat Genet 2004;36:1159–1161.
32 Lane AH, Aijaz N, Galvin-Parton P, Lanman J, Mangano R, et al: Mosaic variegated aneuploidy with growth hormone deficiency and congenital heart defects. Am J Med Genet 2002; 110:273–277.
33 Suijkerbuijk SJ, van Osch MH, Bos FL, Hanks S, Rahman N, et al: Molecular causes for BUBR1 dysfunction in the human cancer predisposition syndrome mosaic variegated aneuploidy. Cancer Res 2010;70:4891–4900.
34 Baker DJ, Jin F, van Deursen JM: The yin and yang of the Cdkn2a locus in senescence and aging. Cell Cycle 2008; 7:2795–2802.
35 Hartman TK, Wengenack TM, Poduslo JF, van Deursen JM: Mutant mice with small amounts of BubR1 display accelerated age-related gliosis. Neurobiol Aging 2007;28:921–927.
36 Matsumoto T, Baker DJ, d'Uscio LV, Mozammel G, Katusic ZS, et al: Aging-associated vascular phenotype in mutant mice with low levels of BubR1. Stroke 2007;38:1050–1056.
37 Coppe JP, Rodier F, Patil CK, Freund A, Desprez PY, et al: Tumor suppressor and aging biomarker p16(INK4a) induces cellular senescence without the associated inflammatory secretory phenotype. J Biol Chem 2011;286: 36396–36403.
38 Naylor RM, Baker DJ, van Deursen JM: Senescent cells: a novel therapeutic target for aging and age-related diseases. Clin Pharmacol Ther 2013;93:105–116.
39 Jeyapalan JC, Ferreira M, Sedivy JM, Herbig U: Accumulation of senescent cells in mitotic tissue of aging primates. Mech Ageing Dev 2007;128:36–44.
40 Rodier F, Campisi J: Four faces of cellular senescence. J Cell Biol 2011;192: 547–556.
41 Tchkonia T, Zhu Y, van Deursen J, Campisi J, Kirkland JL: Cellular senescence and the senescent secretory phenotype: therapeutic opportunities. J Clin Invest 2013;123:966–972.
42 Baker DJ, Perez-Terzic C, Jin F, Pitel KS, Niederlander NJ, et al: Opposing roles for p16Ink4a and p19Arf in senescence and ageing caused by BubR1 insufficiency. Nat Cell Biol 2008;10:825–836.
43 Baker DJ, Wijshake T, Tchkonia T, LeBrasseur NK, Childs BG, et al: Clearance of p16Ink4a-positive senescent cells delays ageing-associated disorders. Nature 2011;479:232–236.
44 Begus-Nahrmann Y, Lechel A, Obenauf AC, Nalapareddy K, Peit E, et al: p53 deletion impairs clearance of chromosomal-instable stem cells in aging telomere-dysfunctional mice. Nat Genet 2009;41:1138–1143.

Vasily Romanov, PhD
Leibniz Institute for Age Research – Fritz Lipmann Institute e.V. (FLI)
Beutenbergstrasse 11
DE–07745 Jena (Germany)
E-Mail vsromanov@fli-leibniz.de

Zhenyu Ju, PhD
Hangzhou Normal University (Science Park)
Wen Yi Xi Lu #1378, D-611
Cang Qian, Hangzhou, Zhejiang 311121 (China)
E-Mail zhenyuju@163.com

Rudolph KL (ed): Adult Stem Cells in Aging, Diseases and Cancer.
Else Kröner-Fresenius Symp. Basel, Karger, 2015, vol 5, pp 83–84 (DOI: 10.1159/000369265)

Thomas Hell Plays 'Vertigo' by György Ligeti

During the social hours of the symposium, we had a piano concert in the historical Festsaal of the Wartburg Castle (UNESCO World Heritage since 1999), one of the best-preserved medieval castles in Germany. Because of its excellent acoustics, King Ludwig II of Bavaria rebuilt the concert hall in the Neuschwanstein Castle. We take this opportunity to thank the world famous pianist Thomas Hell for his poignant and outstanding performance. Thanks to Thomas Hell and Schott Music & Media GmbH, one of his masterpieces performed at the Wartburg concert is available for free download online. It is the étude 'Vertigo' from his recent complete recording of the études of György Ligeti, who would have turned 90 on May 28th, 2013. The access is limited until October 31st, 2015. http://www.fli-leibniz.de/else_kroener_2013.php. PIN code: ekf15r.

Fig. 1. György Ligeti: Études pour piano, deuxième livre, Étude 9, Vertigo = Track 9: WERGO-CD WER 67632, 3:02. Thomas Hell, piano. With permission from WERGO/Schott Music & Media, Mainz. www.wergo.de.

Thomas Hell

> His playing felicitously combines intelligence with virtuosity.
> *Alfred Brendel*

The pianist Thomas Hell has established his reputation in an unusually broad repertoire. His musical explorations include the standard masterpieces, but he is particularly known for his penetrating performances of works by 20th-century composers.

Thomas Hell has performed and recorded major works by Elliott Carter, Charles Ives, Arnold Schoenberg, Luigi Dallapiccola and Pierre Boulez. More recently, Thomas Hell has received great acclaim for his performances of the complete *Études pour piano* by György Ligeti in Darmstadt, Basel and Tokyo. His performance in Tokyo was selected as 'Best Concert 2010' by the Japanese journal *Ongaku no tomo*. In the same year, his CD with piano music by Eduard Steuermann, a pupil of both Busoni and Schoenberg, received the coveted German Record Critics' Award ('Jahrespreis der deutschen Schallplattenkritik' 2010). Thomas' discography also includes piano works by Max Reger and Robert Schumann as well as a disc containing the two violin sonatas by Béla Bartók, performed together with Adrian Adlam, and recently a recording of the complete *Études pour piano* by Ligeti (WERGO).

Thomas Hell has won prizes at several German and international competitions, including the first prize at the Concours international de piano d'Orléans. He has appeared widely throughout Europe as well as in Japan and Russia, and he is regularly invited to perform at major international festivals and in renowned concert halls.

Already as a student at the Hanover University of Music, Drama and Media, Thomas Hell enjoyed approaching music from different angles. He therefore obtained his concert diploma for piano whilst studying with David Wilde, but also earned a diploma in Music Theory, studying with the composer Reinhard Febel. Thomas Hell now teaches at the State University of Music and Performing Arts Stuttgart (Staatliche Hochschule für Musik und darstellende Kunst) and holds master classes at the Kunitachi College of Music in Tokyo, the Tokyo College of Music, the Danish Royal Academy of Music Aarhus as well as the Iceland Academy of the Arts, Reykjavík. More information about Thomas Hell can be found at www.thomashell.de.

Author Index

Subject Index